GAMES PRIMATES PLAY

ALSO BY Dario Maestripieri

Macachiavellian Intelligence:
How Rhesus Macaques and Humans
Have Conquered the World

GAMES
PRIMATES
PLAY

{ *An Undercover Investigation
of the Evolution and Economics
of Human Relationships* }

DARIO MAESTRIPIERI

BASIC BOOKS
A Member of the Perseus Books Group
New York

Published by Basic Books,
A Member of the Perseus Books Group

Designed by Pauline Brown
Typeset in 12 point Goudy Std by the Perseus Books Group

Library of Congress Cataloging-in-Publication Data

Maestripieri, Dario.
    Games primates play : an undercover investigation of the evolution and economics of human relationships / Dario Maestripieri.
        p.   cm.
    Includes bibliographical references and index.
    ISBN 978-0-465-02078-2 (hardcover : alk. paper)—ISBN 978-0-465-02930-3 (e-book) 1. Interpersonal relations. 2. Dominance (Psychology) 3. Control (Psychology) I. Title.
    HM1106.M333 2012
    155.7—dc23

                                                                2011045523

                    10 9 8 7 6 5 4 3 2 1

For my children,
Elena, Luca, and Sarah

CONTENTS

Introduction   ix

Chapter 1:   Dilemmas in the Elevator   1

Chapter 2:   The Obsession with Dominance   17

Chapter 3:   We Are All Mafiosi   53

Chapter 4:   Climbing the Ladder   79

Chapter 5:   Cooperate in the Spotlight,
             Compete in the Dark   109

Chapter 6:   The Economics and
             Evolutionary Biology of Love   143

Chapter 7:   Testing the Bond   171

Chapter 8:   Shopping for Partners in
             the Biological Market   195

Chapter 9:   The Evolution of
             Human Social Behavior   229

             Epilogue   267

Acknowledgments   275
Notes   277
References   285
Index   297

INTRODUCTION

In the Academy Award–nominated 2009 film *Up in the Air*, the protagonist, Ryan Bingham—played by actor George Clooney—is a corporate layoff expert who travels from city to city, firing employees for corporations that are downsizing their personnel owing to the bad economy. Bingham spends his life on airplanes and in airports around the country. He never checks any luggage and fits all he needs for his travels in a small carry-on bag he packs automatically and wheels around with great ease. Packing light is also his life philosophy. When he's not busy firing people, he gives inspirational lectures in which he tells audiences that life is easier and happier without heavy baggage. People are dragged down, he says, by owning properties and maintaining social relationships. He doesn't own a house, or furniture, or any belongings that don't fit into his bag. He has no wife or girlfriend, no friends, and never sees or talks with his sisters or any other family members. Needless to say, life teaches Bingham a lesson: the happiness of traveling without baggage is an illusion. When he falls in love with a fellow traveler, he experiences the real happiness of love and companionship, and when the relationship ends he feels the keen pain of loss and realizes that being alone isn't so fun after all.

Unlike Ryan Bingham, most people don't live "up in the air" where, by constantly being on the move, they might choose not to have stable relationships with others. Instead, most of us generally maintain lifelong relationships with our parents, siblings, children, and other relatives. We also establish and keep up relationships with our romantic partners, friends, coworkers, and even people we've met only on Facebook. Moreover, many of us have intense and long-lasting social bonds with our dogs, cats, and other pets. According to my University of Chicago colleague John Cacioppo, who wrote a book called *Loneliness: Human Nature and the Need for Social Connection*, we all need good social relationships to live a long, healthy, and happy life.[1] People who don't have good relationships with others might think that they are happy, but generally they are not.

Even when we are alone, relationships play a central role in our lives. For instance, while traveling for business, working out at the gym, or lying in bed with insomnia in the middle of the night, our thoughts revolve around relationships: we remember and revisit past events involving ourselves and other people, plan social strategies, or worry about potential future social failures. And as if what goes on in our own relationships were not enough to keep us busy twenty-four hours a day, we also gossip about the relationships of other people we know and even enjoy following the relationships of people we don't know on reality television or in *People* magazine. Relationships have a pervasive influence on all aspects of our lives and affect our thoughts, our emotions, and our health virtually from the cradle to the grave.

Human social relationships can be good or bad, strong or weak, symmetrical or asymmetrical, and everything in between. The characteristics of a relationship between two people are by no means solely the result of their unique personalities, the history of their previous interactions, or the context in which their relationship takes place. Relationships have a life of their own: they begin in a certain way, develop along a certain trajectory, get stronger or weaker over time, and then stabilize or end in predictable ways. Whether they are parent-child relationships, sibling relationships, same-sex or opposite-sex friendships, romantic relationships with or without children, professional relationships, or competitive relationships, all relationships have their own distinctive patterns.

In his best-selling 1964 book *Games People Play*, psychiatrist Eric Berne makes the case that when people interact with their family members, friends, coworkers, or strangers, they do so according to specific patterns that are governed by particular rules and usually characterized by predictable outcomes.[2] Calling these patterns "games," Berne points out that the predictability of these patterns and outcomes stems from our tendency to assume particular social roles in relationships (for example, "the Child," "the Parent," or "the Adult") and that these roles are associated with certain behaviors. Therefore, relationships that involve the same role pairs—such as the Child and the Parent—have a lot in common.

Not surprisingly, our understanding of human relationships has evolved significantly in the half-century since Berne's book was published. Research in psychology and psychiatry has shown that our behavior in social relationships is the result of complex interactions between our genes and our environment and the effects of these interactions on our brains, emotions, and thoughts. In analyzing the complexities of human relationships with increasing (sometimes microscopic) precision, however, researchers appear to have lost interest in their general underlying patterns. They no longer ask why these patterns exist or where they come from. To answer these questions—indeed, to identify the patterns at all—we must step out of the laboratory and take a good look at people and their relationships in the context of other life forms and their behaviors. In other words, we need to venture out of psychology and into biology. Why? Because many of the rules and patterns underlying human relationships developed through evolutionary processes, and those same evolutionary processes have produced similar patterns in other animal species.

As an evolutionary biologist who has studied animal social behavior for almost thirty years, I can attest to the fact that many of the games played by people are also played by other animals. And you don't have to take my word for it. Tens of thousands of studies of animal social behavior conducted in myriad different species during my lifetime—I was born in 1964, the year *Games People Play* was published—have shown that all social animals have relationships with members of their own species. These relationships may be few or many in number, and they may be simple or complex, depending, among other things, on whether the animals live in small or large groups, whether they have a short or long life span, and whether they have a small or large brain relative to the size of their body. Humans are more similar in these characteristics to closely related primates—such as chimpanzees and gorillas, or even macaques and baboons—than to other animals. Therefore, human relationships have a lot more in common with the relationships of these primates than with those of other animals.[3] In short, the games we play with each other are not unique to our species. Other primate species play the same games, or very similar ones. These games were invented neither by us nor by any of the other primate species. Rather, our shared primate ancestors had been playing these games long before the appearance of *Homo sapiens* on this planet. Thus, in order to fully understand human relationships, we must first understand that human nature is a particular, specialized version of a more general primate nature.

So what is this primate nature exactly?

Our Primate Nature

Contrary to popular thought, human beings are by no means the most complex life forms on planet Earth. Evolution by natural selection has produced organisms that far exceed our complexity in terms of how their bodies are built, how they function, how they exploit their environment, and how they reproduce. Just think about the fish on the ocean bottom that live under conditions of total darkness and extreme pressure, or the hermaphrodite worms that have both male and female sexual organs and adapt accordingly, depending on the sex of their partner. As evolutionary biologist Stephen Jay Gould wrote in his 1990 book *Wonderful Life*, every now and then complex organisms go extinct by chance or "bad luck," and complexity alone does not guarantee that a species will survive or be successful.[4] In fact, sometimes complexity can be an evolutionary liability.

Humans are special, however, in one important feature: our brains. Our brains are larger, relative to our body size, and more complex than those of all other organisms. As a result, our mental abilities (for example, our capacity for abstract thinking or complex computations) are far superior to those of other living things. Yet the increase in brain power in our species is no isolated phenomenon but rather part of an evolutionary trend that began in the primate lineage long before the appearance of *Homo sapiens* on this planet.[5] In fact, other species of monkeys and apes with which we share common ancestors in a not-so-distant evolutionary past also show brains of greater size and complexity compared to those of most other animals.

In theory, this evolutionary trend toward increasing brain size in primates, which culminated in the appearance of a very intelligent species, could have occurred in any animal taxon—in insects, reptiles, birds, or some other kinds of mammals. If this had been the case, the earth today would be overpopulated and arguably "ruled" not by people but by gigantic and super-intelligent cockroaches, for instance, or by Godzilla-sized reptiles, or by talking parrots, cats, or dogs. These animals happen to be quite different from primates in many aspects of their lifestyles: how long they live, how they reproduce, what food they eat, and the kinds of societies in which they live. If people had evolved not from primates but from insects, dinosaurs, birds, or other mammals, human societies would be quite different from the way they are now, and the way humans think and behave toward one another would be quite different as well. For example, if humans were an intelligent type of parrot, pair-bonds between men and women would be stronger than they currently are (divorce rates would plummet), women would lay eggs in a nest, and men would help feed chicks by regurgitating food into their mouths; moreover, although we would live in large groups with other birds, there would be no struggles for power involving violence or murder. By contrast, if humans were super-intelligent dinosaurs, pair-bonds would not exist between adult men and women, parental care would be minimal, and we wouldn't live in large and highly structured societies. Social bonds and cooperation between adults would be minimal, individuals would be pretty much on their own, and everyone would be aggressive and dangerous. In a world of large-brained T. rex, life would probably be stressful, with fighting every day, and we'd never know whether tomorrow we were going to eat or be eaten.

Naturally, in these imaginary "human" societies, there would be scientists and philosophers who ask questions about human nature and behavior. The parrots (or dinosaurs) with advanced degrees in evolutionary biology would argue that since humans are birds (or reptiles), studying and understanding birds (or reptiles) in general is a necessary prerequisite for understanding human nature and human behavior. In contrast, monkeys and apes would probably be kept at home as pets or in zoos, studied in the wild as zoological curiosities, or maybe even served as exotic specialties in restaurants. But nobody would write books about "the monkey in the mirror" or "the naked ape" or "our inner chimpanzee."

The way things have turned out, humans happen to be very intelligent primates, not bugs, lizards, birds, or dogs. We share many more of our biological characteristics with other primates than with other animals. Not only is it useful to study primate behavior in general in order to understand our own behavior in particular, but it is especially important to understand the characteristics and behavior of the primates that are most closely related to us, such as apes and Old World monkeys.

Primates are not that different from other animals in their goals of basic survival and reproduction. They need to find and eat food, avoid being eaten by predators, and mate with members of the same species to reproduce. Many of the problems that arise for primates in pursuing these goals and most of the solutions are the same as those for other animals. One important difference, however, is that success in survival and reproduction for primates depends to a much larger extent on the behavior of conspecifics than it does in most other animals. This is especially true for apes and Old World monkeys. One key characteristic of apes and Old World monkeys, which happens to be strictly associated with their intelligence, is their sociality.

A brief comparison of the nature of sociality in chimpanzees with that of other animals illustrates the importance of this point. Many animals, including insects, fish, birds, and nonprimate mammals, live in groups with other members of their species. Daily activities, such as traveling, feeding, or sleeping, take place in close proximity to other individuals. Members of the group, however, don't necessarily compete with one another for food, shelter, or attractive mates, and there is often little or no need for cooperation between group members. In these animal societies, individuals by and large mind their own business and don't become entangled with "friends" or "enemies." The disappearance or death of an individual is largely inconsequential for the rest of the group and may even go unnoticed. The life of a chimpanzee, in contrast, is intertwined with the lives of all the other chimpanzees in the group, forming a thick web of intricate connections. Every move a chimpanzee makes on the social chessboard has an effect on every other chimpanzee's life, whether they like it or not. This has many implications for the behavior of individuals and for what it takes to be socially successful in a chimpanzee group.

Chimpanzees and humans live in highly competitive societies. Instead of fighting all the time, individuals establish dominance hierarchies within their group. High-ranking individuals have preferred access to food, shelter, and attractive mating partners. In addition to the difficulties in finding food and mates and their exposure to more risks, low-ranking chimpanzees are also chronically stressed by aggression and intimidation from above. As a result, low-ranking individuals are more likely to be in poor health, to die younger, and to leave fewer descendants than their high-ranking brethren. To attain high social status, chimpanzees must form alliances with other individuals and receive their support. For example, chimpanzee males form alliances with their brothers and occasionally with unrelated but powerful adult males to win fights against other group members. Competition and cooperation with other group members are pervasive aspects of the social lives of chimpanzees, other apes, and Old World monkeys—and the social lives of humans—to an extent that, with a few exceptions, is not seen in other animals.[6]

Games Primates Play

An underlying theme of *Games Primates Play* is that human nature is manifested in our social interactions more than in any other aspect of our behavior or intellectual activity. This has two major implications. First, since our social behavior has been strongly shaped by evolutionary processes such as natural and sexual selection, we can explain it using cost-benefit analyses and other rational models of behavior (for example, game theory) developed by evolutionary biologists and behavioral economists. Second, the same selective pressures from the social environment that shaped our behavior and that of our primate ancestors may have shaped the behavior of other extant primate species and their ancestors. Therefore, there may be important similarities in social behavior between ourselves and other primate species because we have adapted to similar social environments. It is also possible that natural selection acted on the behavior of the ancestors we share with other primate species and that both humans and modern primates have inherited some aspects of their social behavior directly from their common ancestors. Therefore, other important behavioral similarities between ourselves and other primate species may be due to our common ancestry. In *Games Primates Play*, I examine human social behavior with both rational scientific models and evolutionary and comparative arguments, using examples from closely related primates that live in societies similar to our own.

Others before me have taken this approach to try to elucidate aspects of human nature and behavior. Evolutionary psychologists have shown that humans possess social propensities that have been shaped by natural and sexual selection. For example, the differences in the characteristics that men and women find attractive in a potential long-term romantic partner are likely to be the product of sexual selection.[7] Similarly, economists have established that the choices we make in a variety of not only financial but also social circumstances can be explained by rational models based on the assumption that we maximize the benefits and minimize the costs of such decisions. *Freakonomics*, the best-selling book by Steven Levitt and Stephen Dubner, offers examples of this approach.[8] Finally, primatologists have proved that humans are similar to chimpanzees in that human males, like chimpanzee males, are generally more physically aggressive and violent than females; as eloquently discussed by Richard Wrangham and Dale Peterson in their 1996 book *Demonic Males: Apes and the Origins of Human Violence*, it is likely that both humans and chimpanzees inherited this sex difference in aggressiveness from their common primate ancestors.[9]

What I aim to do in this book is to show that the adaptiveness of our behavior and its evolutionary legacy extend to both the most mundane and the most specialized aspects of modern social life. We often assume that our behavior in everyday situations simply reflects our unique personalities, the choices we freely make, or the influence of our environment. In reality, people around the world, living in very different environments and exposed to very different cultures, behave the same way in these situations. We don't recognize these similarities in part because we are often unaware of our own behavior and in part because we don't pay too much attention to what others do.

For the past twenty years, I have observed human beings in all sorts of social situations, applying to my own species the scientific rigor with which I study other primates. I have gone undercover to report on tics and strange ritual behaviors practiced—often without knowing why—by my species and to examine the curious unspoken customs that govern our behavior in public and in private. I have looked specifically for those behavior patterns and customs that would seem to be governed by nothing but free will but that are so similar from person to person that they reveal much more than individual choices at work. These behaviors, the legacies of our primate past, do not lie hidden—they play out on the surface of our lives, yet are so instinctual, so "natural" to us that we don't notice them. But I have noticed.

To detect the "games" people play in everyday social interactions, it is necessary to become an excellent detective: one must observe human interaction not only closely but without being too overt or obvious. To understand the rules that govern primate games, it is also necessary to know the scientific principles that ethologists, psychologists, economists, and other behavioral scientists have discovered in their quest to unravel the complexities of behavior. Armed with these skills and principles, I have tried to show how the primate past influences our decisions and actions in ways we often do not perceive or understand.

We may think we have outgrown the conditions that govern the lives of other primates. We no longer live in the jungle and swing between trees; instead, our homes are in or around large cities, and we drive cars, wear clothes, spend years in formal education, and communicate electronically. Yet technology and clothes cannot disguise the inheritance of our primate past. They have simply changed the arena in which we act out age-old rituals, making the games that human primates play more arbitrary perhaps, but no less powerful.[10]

Chapter 1

Dilemmas in
the Elevator

The Cavemen's Legacy

In one of the scariest scenes in Brian De Palma's 1980 film *Dressed to Kill*, Kate Miller (played by actress Angie Dickinson) is in the elevator, on her way up to the seventh floor. When the elevator stops and the door opens, the killer—a man wearing a woman's wig, dark sunglasses, and a black coat—walks in, a razor in his hand. Kate raises her hand to protect her face, but the killer slashes it with the razor blade and continues to hack away at her until the elevator reaches the ground floor, where the door reopens and the two people who have called the elevator see Kate's body on the floor, covered with blood.

In the movies, more people are probably murdered in elevators than in any other closed space—perhaps with the exception of the shower. In reality, the probability of being the victim of a deadly attack in an elevator is virtually zero. Yet the way people act toward others when they ride together in an elevator suggests that they have serious concerns about their safety. If the elevator is crowded, everybody stands still and stares at the ceiling, the floor, their watch, or the button panel as if they've never seen any of these items before. When two strangers ride together, they stand as far apart as possible and avoid facing each other directly, making eye contact, or making any sudden movements or noises.

You might think that strangers in an elevator are simply trying to be polite in a socially awkward situation, but the truth is that much of our elevator behavior is not the result of rational thinking. It's an automatic, instinctive response to the situation. The threat of aggression is not real, yet our minds respond as if it were and produce behaviors that are meant to protect us. Elevators are relatively recent inventions, but the social challenges they pose are nothing new. The scenario of being in close proximity to others in a restricted space has been repeated innumerable times in the history of humankind.

Imagine two cavemen of the Paleolithic Age who happen to separately follow the tracks of a large bear into the same small, dark cave. There each discovers not a bear but another hungry caveman ominously waving his club—clearly an awkward situation that requires an exit strategy. In Paleolithic days, murder was an acceptable way to get out of socially awkward situations (the way we use an early morning doctor's appointment today as an excuse to leave a dinner party early). In the cave, one of the cavemen whacks the other over the head with his club and the party's over. Occasionally, the caveman's chance encounter is with a female of the species, which makes it an opportunity for reproduction. But if a male caveman encounters another of his kind, it's bad news. Similarly, when male chimpanzees in Uganda come across a stray male from another group, they slash his throat and rip his testicles off—in case he survives and has any future ambitions for reproduction.

Our minds evolved from the minds of these cavemen, and their minds, in turn, evolved from those of their primate ancestors—apes that looked a lot like chimpanzees. Although some of our mental abilities appeared relatively recently in our evolutionary history—like our capacity for abstract reasoning, language, love, or spirituality—the way our minds respond to potentially dangerous social situations is nothing new. Just as the way we feel pain in response to bodily injuries probably hasn't changed in millions of years, the way primate minds respond to social threats hasn't been modified very much either. On the contrary, evolution has been so conservative in this domain that the minds of humans, chimpanzees, and even macaque monkeys—

whose ancestors began diverging from ours 25 million years ago—still show traces of the original blueprint.

The way people behave in elevators is not a popular topic for scientific research these days, but it was all the rage in the 1960s. An anthropologist named Edward T. Hall wrote a book in 1966 called *The Hidden Dimension*, in which he argued that when a person invades someone else's personal space all kinds of trouble ensue.[1] According to Hall, personal space is like an invisible bubble that people always carry around themselves. The radius of the bubble can be short or long, depending on the individual or the cultural norms of the society in which he or she lives. Noting that human personal space is the equivalent of an animal's territory, Hall suggested that aggressive responses to violations of personal space represent attempts to defend one's territory.

Given what we have learned about animal behavior since Hall wrote his book, the analogy between human personal space and an animal's territory is no longer useful. Territorial behavior is quite rare in primates and mostly confined to species that are only distantly related to humans, such as lemurs or New World monkeys. Moreover, humans do not aggressively defend the invisible bubble around themselves the way territorial animals defend the place where they live. What we do instead is take measures to protect ourselves from the risk of aggression whenever potentially dangerous individuals are close to us. Being next to another individual simply increases the probability of aggression, especially if this individual is a stranger. The relationship between close proximity and risk of aggression has been studied and is well understood in other primates, including species that do not defend territories, such as rhesus macaques and baboons.[2] By recognizing the evolutionary continuity between the human mind and the minds of nonhuman primates, it becomes clear that people's reaction to the presence of others in an elevator is simply a response to the risk of aggression.

The risk of being slashed by a maniacal murderer with a razor blade is not the only problem entailed by close proximity to strangers in restricted spaces. The anxiety associated with the anticipation of

Figure 1.1. *Cartoon courtesy of Jason Love.*

"Just so you know: If this elevator breaks
down, I have no problem cannibalizing
your body for my survival."

danger can be just as bad for our health as any physical injury resulting
from actual aggression. People in elevators sometimes show stress-
related behaviors: they scratch their head even if it's not itching,
they pick the cuticles around their fingernails, or they check their
wristwatch compulsively even though they already know what time
it is. Elevator stress is mild compared to the stress of being robbed at
gunpoint, yet the difference between the two experiences is only a
matter of degree. Just as our mind knows that aggression is dangerous
and is prepared to take steps to avoid it, our mind also knows that
stress is not good for us and is prepared to deal with that as well. This
is true not only for strangers who find themselves alone in an elevator
but also for monkeys trapped inside a small cage.

To Fight or Not to Fight

Imagine this situation. A rhesus macaque, which normally spends its
days roaming in the jungle surrounded by fellow macaques, is suddenly
introduced into a small cage by an undergraduate student eager to

publish his first scientific article about monkey behavior. The ambitious student then introduces another rhesus macaque into the same cage—and waits and watches.

The risk of serious fighting between the monkeys is very high. The rules of macaque society are such that, whenever a monkey is close enough to another monkey that it can quickly grab and bite it, chances are good that this will happen.[3] Furthermore, space restriction prevents the monkeys from running away if an attack is launched. Thus, in a small cage, aggression can be easily triggered—and once it's started it cannot be easily stopped.

If the two monkeys have never met before, the risk of serious fighting rises even higher. Macaque monkeys don't like strangers, so unless the other individual is a potential sexual partner, its presence could immediately elicit a hostile response. Furthermore—as we'll see later—although the two monkeys could work together to reduce the tension, the fact that they don't know each other makes it difficult for them to cooperate. The minds of the two people in the elevator process their situation using the same simple arithmetic: stranger + restricted space = trouble. An emotional alarm goes off immediately, as when we touch a flame with a finger and immediately feel the pain of being burned.

But if the risk of fighting in a restricted space is so high, why don't monkeys—or people—just go ahead and fight?

People and monkeys generally have no trouble fighting with members of their own species, whether they be strangers, friends, or family. The times and places when fights take place, however, are rarely random. Whether we challenge a fellow army officer to a duel in the park or meet the local bully for a fistfight in the dark alley behind the high school, the logistics of the confrontation are carefully chosen in advance. Some benefits of agreeing on such details include opportunities to strike without being seen, to avoid retaliation, to limit the damage if defeated, or to receive support or protection from other individuals.

None of these benefits are available to people stuck in an elevator or to a pair of monkeys isolated in a cage. In these situations, there is no guarantee that victory can be achieved or that the costs of defeat can be controlled. There is a good chance that both parties will lose, and lose big. Among monkeys and Paleolithic cavemen, just as among modern humans, there are always losers who make wrong decisions and pick a fight in the wrong place. Natural selection, however, does not reward them. Over the evolutionary history of our species and that of other primates, the individuals with genetic predispositions for resisting the impulse to fight in the wrong place have had longer lives and produced more offspring than their indiscriminately belligerent counterparts. Consequently, the descendants of these wise individuals have genetically inherited behavioral strategies that allow them to avoid fighting in elevators or in small cages. These behavioral strategies appeared a long time ago in the evolutionary history of the Primate order, and they have worked so well for so long that natural selection has left them almost unchanged in our primate minds.

It turns out that when two rhesus macaques are trapped together in a small cage, they try everything they can to prevent a fight. Moving with caution, acting indifferent, and suppressing any behaviors that could trigger aggression are good short-term solutions to the problem. The monkeys sit in a corner and avoid any random movements; even a brief touch could be interpreted as the beginning of hostile action. Mutual eye contact is also dangerous because, in monkey language, staring is a threat. The monkeys look up in the air or at the ground, or they stare at some imaginary point outside the cage. But as time passes, sitting still and feigning indifference are no longer sufficient strategies to keep the situation under control. Tension builds between the monkeys, and sooner or later one of them will lose its temper. To avoid immediate aggression, and to reduce stress, an act of communication is needed to break the ice and make it clear to the other monkey that no harm is intended (or expected). Macaque monkeys bare their teeth to communicate fear and friendly intentions. If this "bared-teeth display"—the evolutionary precursor to the human smile—is well received, it can function as a prelude to grooming. One monkey brushes and cleans the other's fur, gently massaging the skin while picking and eating parasites. This act can both relax and appease the other monkey, virtually eliminating the chance of an attack.

So, if you are a rhesus macaque and find yourself trapped in a small cage with another macaque, you know what to do: bare your teeth and start grooming.[4] If you are a human and find yourself riding in an elevator with a stranger, in theory you could do the same thing (or the human equivalent thereof): smile and make small talk.

In practice, however, things are usually a little more complicated than that.

Dilemmas in the Elevator

When you walk into a crowded elevator, you may not have many options for action. Typically, you make a 180-degree turn and stand in front of the door with your back to the other people. Everybody else stands still and stares at the ceiling. Walking into an elevator occupied by only one other person is trickier. Should you acknowledge the other person? Feigning indifference is risky—it could be taken personally. Should you smile and say something pleasant? What if your friendly overture is misinterpreted or unwelcome? Should you puff yourself up and stare the stranger down to clarify who would win out in a potential confrontation? What if the other guy gets mad and turns into the Incredible Hulk? These are dilemmas that people living on the lower floors of high-rise buildings must face every morning on their way to work. They walk into the elevator and, chances are, someone else is already there.

Some time ago, I had the privilege of living on the twentieth floor of an apartment building in Chicago, where I not only enjoyed a view of Lake Michigan but also had the perfect opportunity to observe how people behave when they encounter others in a restricted space. Every morning the elevator made at least one stop on its way down from my floor and typically picked up one passenger. The building contained more than two thousand apartments, so almost every day I encountered someone I hadn't previously met. Since observing behavior is what I do for a living, I couldn't help paying attention to how each new stranger acted and filing it in my mental archive.

The following describes a typical interaction. The elevator stops on the fifteenth floor and a man in his thirties, unshaven and wearing sweats, walks into the elevator. The stranger looks at me for a nanosecond, then looks at the button panel. I've already pressed the ground-floor button, and it's the only one that's lit up. Two or three seconds go by—the time it takes him to make a decision. He presses the ground-floor button again, takes one step back, and stands in a corner while continuing to stare at the panel. It's obvious to me that the stranger's behavior is not the result of failure to notice that the button was lit, or simply habit (the same action repeated mechanically every day, without thinking). It's clear that he has looked at the button panel, paused, and then deliberately pressed the ground-floor button again. Why has he done that?

Here's what I think is going on in this situation. The stranger and I are both aware of each other's presence—that's undeniable. By pressing the button again, the stranger is refusing to acknowledge my nature as an intentional being—someone who has a goal (reaching the ground floor) and has already taken action to achieve it (pressing the button). Recognizing that other people have goals and desires and that their behavior is guided by the pursuit of these goals is a complex cognitive ability that arguably sets humans apart from all other animals, including monkeys and apes.[5] This faculty comes to us as part of a mental package that also includes the capacity to empathize with our fellow human beings—to understand their feelings and feel their pain. Attributing goals and feelings to other people is what makes us recognize them as humans—a hallmark of personhood, if you wish—and we make the same attributions with our beloved pets. By pressing the ground-floor button again, the stranger has failed to acknowledge my status as a fellow human traveler with similar goals and engaged in goal-directed actions. His behavior dehumanizes me and, from his perspective, makes me safer and easier to deal with.

Another morning, another elevator ride, and another stranger joins me on the way down. This time it's a gray-haired man wearing a business suit and toting a black leather briefcase. The man notes that the ground-floor button has already been pressed and does not press it again. He stands quietly across the elevator from me, his gaze going up and down along the edge of a wall panel. By refraining from pressing the button again, this stranger has implicitly acknowledged my presence and my previous actions. We are sharing not only the space within the elevator but also our goals. We are traveling to the ground floor together. Sharing goals is a crucial aspect of a social relationship. Parent-child bonds, friendship, and romantic love are all based on an acknowledgment that two individuals have the same goals and are willing to engage in actions to jointly pursue them. The stranger's behavior is not an act of friendship or love, but it's an important move that signals a friendly disposition.

Other aspects of the stranger's behavior suggest that he is concerned about the situation, like anyone else would be. Staring at an imaginary point in space is an unnatural behavior that might attract another individual's attention and get on his nerves. Horizontal gaze movements are dangerous in an elevator because they inevitably result in eye contact, unless they occur below the waistline—dangerous for other reasons. Vertical gaze movements along wall corners or door edges, however, like those of the man with the briefcase, are safe because they minimize the chance of having someone's face in the visual field. If the stranger is feeling particularly tense that day, he might even smile at me. On a long elevator ride, however, indifference and awkward smiles would not be enough to keep the peace: the two people would have to start grooming. In other words, they would have to talk.

Games Monkeys Play

What determines whether two individuals respond to each other with hostility, indifference, or friendliness when they meet in a potentially risky situation like being trapped in a closed space? According to Shelley Taylor, a social psychologist at the University of California–Los Angeles and author of the book *The Tending Instinct*, gender makes a difference.[6] Taylor proposes that males show a "fight-or-flight" response to social stress: they either run away, to avoid the stressor, or stay and fight. Females, on the other hand, "tend and befriend": they stay put and behave nicely to try to win over the enemy.

Taylor is probably right. If two male macaque monkeys are trapped together in a cage with no opportunity for escape, there is a good chance that they will kill each other. Two female macaques in the same situation might instead try to be nice to each other and work together to diffuse the tension. However, this is what males and females may do *on average*—not all males and females behave in full conformity to Taylor's hypothesis. In reality, the line that demarcates the male and female strategies is crossed all the time—in both directions.

The desire to gain a deeper understanding of the monkey mind—as well as to publish my first article—was what prompted me to design an experimental study of monkey behavior at the University of Rome many years ago. I was the originator of the aforementioned "elevator" experiment: having placed two macaques together in a small cage, where they barely had room to stand up and turn around, I videotaped their behavior for one hour. I used only females for the experiment because I was afraid that two male monkeys would fight and kill each other, and also because we had many more adult females than adult males in the lab. I tested over twenty-five pairs of monkeys. In about half the tests, the two monkeys knew each other, although they had lived apart for many months prior to the experiment. The other pairs were made up of individuals who had not met or seen each other before.[7]

When the monkeys who knew each other met in the cage, they initially looked uncomfortable but quickly figured out how to defuse the situation. They started grooming each other and continued to do so for most of the hour. They took turns so that by the end of the hour, each had given and received a similar amount of massaging. The continuous grooming reduced the tension between the monkeys and eliminated any risk of conflict—in the end, everyone seemed happy.

When the strangers were paired together, however, the tension in the cage was so thick that you could cut it with a knife. The monkeys glanced nervously in all directions and scratched themselves like crazy—a sign of anxiety in macaques. Although they took extra care to avoid any eye-to-eye contact, it looked as if they were surreptitiously sizing each other up and thinking, *Is she bigger than I am? Is she mean? Can she really kick my butt?* This would go on for several minutes. Then, in some pairs, one of the two monkeys—the one that presumably answered *Yes* to all three questions—would freak out and "smile" submissively to the other. She would start grooming while the other sat and enjoyed it. When the groomer's fingers were so sore that she had to take a break, the other female would return the favor for a few seconds, but would then immediately stop and get comfortable for the next session. In the end, these unfamiliar pairs groomed almost as much as the familiar pairs, but the relationship was decidedly lopsided: one of the females did all the work, while the other simply reaped the benefits of the grooming.

There were also pairs of unfamiliar monkeys—about half of the total—in which neither monkey acted as if she was intimidated by the other. No submissive smiles were exchanged. Nothing happened for several minutes until, finally, one of them started grooming the other. This monkey stopped after a few seconds, however, and immediately lay down in front of the other or placed her leg, arm, or butt right in the other's face—a request for grooming. The other monkey, in turn, did exactly the same thing: groomed for a few seconds and then requested reciprocation. Some didn't bother to groom at all and simply asked for more grooming. The two monkeys played the game "you groom me–no you groom me" over and over, with the result that by the end of the hour they had exchanged little actual grooming. They appeared to be as uncomfortable and anxious as when the test started.

I was initially puzzled by these differences in behavior among the monkey pairs but became excited when I discovered that the monkeys' behavior was perfectly explained by a branch of economics called game theory. Without being conscious of it, my monkeys were playing a game known among economists as the Prisoner's Dilemma, which explains the exchange of altruistic behavior between two unrelated individuals.[8]

This model was originally developed in 1950 by two American mathematicians, Merrill Flood and Melvin Dresher, while they were working for the Rand Corporation, and it was later formalized by another mathematician, Albert Tucker, who coined its name. The game is illustrated by this scenario. Two prisoners are interrogated in separate rooms for a crime they committed together. They are not allowed to communicate with each other. If both prisoners remain loyal and refuse to incriminate the other, they will each get a mild sentence of one year in jail. If one confesses and incriminates the other, the confessor will walk free while the other receives a sentence of five years. However, if each incriminates the other, both prisoners will be sentenced to three years. The situation can be thought of as a game in which the two prisoners are the players and the years of the sentence are the payoffs. The game has two possible strategies: cooperating or defecting. As illustrated by the payoff matrix shown in Figure 1.2, the player who defects always gets a lighter sentence, no matter what the other one does. When player 1 defects, he gets a sentence of zero years if player 2 cooperates, and three years if player 2 defects. In contrast, when player 1 cooperates, he gets a sentence of one or five years, depending on player 2's behavior. Although defecting is generally the best strategy, if both players cooperate they receive higher payoffs than if they both defect. Cooperating, therefore, is a winning strategy, but only when a player is certain that the other player will cooperate as well.

When the game is played only once—that is, each player is allowed only one move—and with a stranger, the chances of cooperation are slim. Since the two players don't know each other, they have no reason to expect cooperation from the other. Moreover, since the game will not be played again, it is pointless to cooperate and expect future reciprocation. In this case, the best strategy is to defect. However, when the game is played repeatedly—also known as an "iterated Prisoner's Dilemma"—there is an opportunity to keep track of the other player's previous moves and to act accordingly. Computer simulations conducted in the late 1970s by Robert Axel-

Figure 1.2. *Example of a payoff matrix for the Prisoner's Dilemma.*

		Player 2	
		Cooperate	Defect
Player 1	Cooperate	1 year	5 years
	Defect	0 years	3 years

rod, a political scientist and author of the book *The Evolution of Co-operation*, showed that under these circumstances the winning strategy is neither cooperating nor defecting, but a new strategy called "tit-for-tat." In this strategy, player 1's first move is to cooperate, and then to simply copy what player 2 has done in his previous move; if player 2 cooperates, player 1 cooperates as well. If player 2 defects, player 1 defects too. Axelrod suggested that tit-for-tat has three characteristics that make it a winning strategy: *niceness* (the tit-for-tat player is never the first to defect); *retaliation* (the tit-for-tat player is no fool and immediately retaliates against defection with another defection); and *forgiveness* (the tit-for-tat player remembers only one move back in time and "forgives" a player who defected in the past if his most recent move is to cooperate).

The dynamics of the Prisoner's Dilemma can be significantly altered by two factors. One of them is kinship. If two individuals are family members, they may be willing to behave altruistically without expecting reciprocation. For example, a monkey mother would be happy to groom her young daughter for hours without obtaining any grooming in return. Monkeys and humans are happy to help relatives because they share their genes, and by behaving altruistically they increase the chances that their own genes will be maintained in the population. In addition to kinship, the dominance relationship between the two players can alter the dynamics of the Prisoner's Dilemma. In this case, the individual who is subordinate is willing to behave altruistically toward the dominant individual, not in order to be reciprocated in the same currency, but in exchange for safety or protection. I explore this issue in more depth in Chapter 2.

My elevator experiment created a risky situation for the monkeys involved. The circumstances called for action to reduce the risk of aggression and to alleviate tension, and as we've seen, monkeys are predisposed to use grooming behavior to handle these situations. Grooming happens also to be an altruistic behavior that benefits the recipient and entails a cost in time and energy to the giver. The dynamics of grooming in the various experimental pairs demonstrated that the Prisoner's Dilemma is a powerful model to explain the exchange of altruism, not only among people but among monkeys as well. All the monkeys in my experiment were unrelated to one another. If the two monkeys in the cage had met before, they responded to the situation as if they expected to meet again in the future. Both individuals in the familiar pairs played the cooperation strategy and got the most out of the situation. The monkeys in the unfamiliar pairs, however, acted as though they were playing the Prisoner's Dilemma game only once: they had met the other player for the first time and had no reason to expect that they would meet again. In half of the pairs, there was a strong asymmetry in perceived power. The female who felt most vulnerable cooperated, presumably in hopes of exchanging grooming for safety. The other monkey took advantage of the situation and defected. In the other half of the pairs, in which there was no clear asymmetry in power or perceived vulnerability, both prisoners played tit-for-tat and ended up retaliating against each other's defections or exchanging small amounts of grooming. At the end of the test, these monkeys looked stressed out—that hour together must have felt like an eternity.

Verbal Grooming

Riding in an elevator with someone else is not usually so stressful as to cause a heart attack. Although an instinctual alarm goes off in response to a perceived risk of aggression, the elevator ride is brief enough that simple indifference is an effective strategy to ensure safety. If elevator rides were one hour long, as in my monkey experiment, I would expect people to use social strategies to reduce the tension, such as smiling and making polite conversation. Insofar as these strategies require cooperation between two individuals, I would expect that the interactions between them would unfold according to the Prisoner's Dilemma dynamics illustrated by the monkey experiment. Some of these dynamics can indeed be observed in everyday elevator interactions.

One morning, for example, I rode the elevator with a middle-aged man who seemed to be particularly intimidated by my presence. As I stepped in, he smiled nervously and started talking immediately. He talked nonstop and managed to relate his entire medical history, complete with symptoms, diagnoses, and treatments, before we reached the ground floor. I doubt that this man thought I was a doctor or that he expected to receive medical advice. Rather, he was more likely an insecure and emotionally vulnerable person who used massive verbal grooming to appease a perceived potential aggressor in a risky situation.

Not all my experiences are like this, of course. When I ride in an elevator with an attractive woman, I'm generally treated with indifference, and I have a hard time believing that response stems from fear or intimidation. When my girlfriend rides in an elevator with a man, he will often strike up a conversation with her and end up asking for her phone number. People's responses to potential mating opportunities are just as predictable as their responses to potentially dangerous situations.

The beauty of human nature, however, is that although people's average behavior can be scientifically predicted, there is a lot of unpredictable variation above and below the mean. Once, on the way up to my apartment, an old lady got in the elevator on the second floor, pressed all the buttons for the third through the twenty-second floors, and walked out on the third floor with a grin on her face.

Chapter 2

The Obsession
with Dominance

An Ancient Use of a New Technology

When I wake up in the morning, my brain immediately needs coffee and my body needs sugar. After my biochemical needs have been satisfied, I'm ready to check my email. I log in to my account and find the inbox filled with new messages from family members, close friends, friends I haven't heard from in years, coworkers, business-men from Nigeria, and other strangers with names I can't even pro-nounce. I feel overwhelmed. Email, as anyone with an Internet connection knows, makes life easier but can also be a major source of psychological stress. I start deleting unread messages. The first ones to go are those from people with obviously fake names announcing an unexpected—and unlikely—inheritance or lottery win. More authentic-seeming messages from strangers will be read later. I turn my attention to emails from people I know. I'm eager to read and reply to a few. Some of my responses are long and personal, others short and professional. I choose not to reply to some, letting them stew in my inbox for days until I'm overwhelmed by guilt. Then I write to some people who probably don't expect to hear from me. Just as some people want something from me, I want something from others. As a good citizen of the global community, I make my own contribution to the clogging of inboxes around the world.

It's hard to deny that email makes our work lives significantly eas-ier and more efficient. Email, however, is not only about work. The truth is that, like most other people, I maintain social relationships via email. I exchange emails with my mother and sister in Italy, with friends all over the world, with colleagues and students, and with other people I know.

As someone who studies the social relationships of human beings and those of other primates, I often wonder whether the use of email and all of its derivatives—Facebook, Twitter, and Google+, to name a few—has altered human social relationships in some fundamental way. We humans have evolved to interact with others face-to-face. For millions of years before speech evolved, early humans and their primate ancestors maintained social relationships only with individuals they could see, hear, and touch. To negotiate the day-to-day problems of these social relationships, they used facial expressions, vocalizations, and a great deal of touching, grooming, hugging, and occasionally throwing a slap or a punch. Eye contact is important to nonhuman primates, and also to modern humans, as a gauge of whether another individual is friendly or hostile, dominant or subordinate, sexually attracted or not. But what has happened to all of these measures now that we negotiate social relationships through our computers? Social media users have found creative ways to communicate their moods with "emoticons"—smiling faces, winks, and frowns—to compensate for the fact that the reader may have no idea whether we are serious or joking when we say such things as: "Sometimes you really make me want to kill you." Is that it? Emoticons do it all? The legacy of millions of years of negotiating social relationships face-to-face disappeared instantly the day Al Gore invented the Internet?

I don't think so. Although our high-tech way of communicating might seem to preclude a strong influence of our evolutionary past on the way we act, the rules regulating primate relationships resurface even when we sit down at our keyboards to catch up with friends or reply to work memos. For example, the concern with social status that characterizes the relationships of other primates such as macaques, baboons, and chimpanzees has not disappeared in cyberspace, but is simply expressed in a new and different form.

There are some clear patterns in the way we use email. First of all, email communication between people who know each other well occurs in "conversation" bouts in which several messages are exchanged back and forth over the course of minutes, hours, or a few days. Who starts and who ends the conversation, the time taken to reply, and the length of the transmissions are not random. Let me illustrate this point by taking as an example an email exchange with an imaginary graduate student in my research group whom I'll call Jennifer. One day Jennifer—who has not seen me around for a while because I've been hiding in a coffee shop, trying to write a book without being interrupted—begins an email exchange by sending me a long message. It contains numerous questions and requests for information and for action (there is always something my students need from me). It's quite clear that a response is needed with some urgency. Jennifer obviously put a lot of effort into writing this email—time, energy, and the cognitive resources expended in the production of grammatically correct sentences—but the cost of this initiative, so Jennifer hopes, will be offset by the great benefit that my reply will bring her. By hitting Send and beginning the conversation, Jennifer has made an investment that she hopes will bring a significant return.

From my perspective, writing to Jennifer entails a cost—I am being distracted from my book!—and little to no benefit, if we forget for a moment that I get paid a decent salary for advising students. So I procrastinate in responding to her email, and when I do finally respond, I compose a brief missive in which I provide all the requested information while keeping the word count as low as possible. Jennifer finds my reply encouraging—the investment is beginning to produce a return—and within seconds of hitting the Reply button, I receive another message from her, as long as the first, with more questions and requests. My conversation with Jennifer continues along this pattern: her emails progressively get quicker and longer, while mine get slower and shorter. After responding five times, I let Jennifer's email number six sit indefinitely in my inbox without a reply. According to the rules and etiquette of email communication, a lack of response ends the conversation. In rare cases, a pushy student will attempt to resume the conversation by sending an email that begins, "I don't know if you received my last email but in case you didn't, this is what I wrote, . . . ," followed by the text of the message I ignored. Jennifer, however, knows better, understands and respects the conventions of email, and patiently waits a few days before starting a new conversation on a different topic. The way emails are exchanged between professors and students is analogous to how it's done in the workplace: similar exchanges occur between bosses and their direct reports. Employees often write unsolicited emails to their superiors, while the latter are not as communicative. They may not respond at all to unsolicited queries, reply with automated messages, or forward the email to their secretaries.

What students and professors have in common with employees and their superiors is a clear dominance relationship—one individual is dominant, the other is subordinate. The costs and benefits of exchanging emails are different for the dominant and subordinate parties, and as a result a distinctive pattern of email exchange emerges. Of course, it's possible that professors and bosses simply have less time for email than students and employees. But I suspect that, at the end of the day, professors and supervisors spend as much time on email as everybody else. They write long and unsolicited emails too, just not to their students or employees. They write instead to someone—possibly a superior—from whom they want or need something. So this asymmetry is not a matter of time. It's dominance: the subordinate writes more, the dominant less.

Consider an exchange of grooming between a dominant and a subordinate male chimpanzee.[1] Imagine that the dominant chimpanzee is sitting in a corner by himself, minding his own business, when the subordinate approaches him, "smiles" a couple of times, and begins to groom him. The subordinate grooms the dominant for a long time, putting a lot of effort into it. It's an investment that is meant to bring returns: grooming, tolerance, or support from the dominant. When the subordinate's fingers become sore, he stops and requests to be groomed back. The dominant doesn't immediately reciprocate; in fact, he doesn't do anything at all until the subordinate,

Figure 2.1. *Male chimpanzees grooming each other in Kibale National Park, Uganda. Photo courtesy of Dr. John Mitani.*

tired of waiting for reciprocation that doesn't occur, resumes his grooming. After a few minutes, he stops again. This time the dominant waits for twenty seconds, then starts grooming the subordinate—but only for a few seconds! Then he stops and waits for the subordinate to get back to it.

The chimpanzees go back and forth like this a few times; their exchange of grooming is lopsided because the dominant takes longer and longer to reciprocate, and the grooming he does give gets shorter and shorter in duration. At some point, the dominant ends the conversation; he stops responding, gets up, and walks away. Recognize the pattern?

What's more, when the dominance relationship between two male chimpanzees is reversed—the subordinate becomes dominant and the dominant becomes subordinate—this change is reflected in their grooming behavior as well. Now it's the former dominant who does all the work, while the other barely reciprocates. Reversals of dominance are uncommon in the human workplace, but it's not unusual for some dominance relationships to become more balanced and symmetrical over time. A former student of mine who, like Jennifer, used to write me long unsolicited emails, has become a professor in a top-notch university. Now when we exchange emails, either one of us is equally likely to start or end the conversation, and he shoots off one-line responses he never would have sent me before. His style of email has changed gradually over time as his career has advanced and he's caught up with me.

Just as examining the grooming behavior of chimpanzees can give us an insight into their social strategies—for example, whether they attempt to increase their social status by befriending powerful individuals or by challenging the authority and privileges of the powerful—the way we use email can tell us something about our own status and potential for advancement. Show me your emails and I will tell you whether you are on a fast track to become a leader of your company, or whether it's unlikely that you will have secretaries answering your emails anytime soon. To understand why dominance affects the exchange of human email and chimpanzee grooming the way it does, let's backtrack for a moment and pick up where we left off in the previous chapter.

Social Relationships and Their Problems

An encounter in a restricted space with a stranger who may be hostile has represented a potentially life-threatening situation over millions of years of primate evolution, so it makes sense that our minds are predisposed to come up with appropriate protective responses. The representative feature of social interaction in our everyday lives, however, is not the chance encounter with a stranger in the elevator, but repeated interactions with people we know well: our family members, romantic partners, friends, and coworkers. We establish and maintain long-term relationships with these people and derive obvious benefits from them.

Although social relationships can be cooperative or competitive, good or bad, they all pose certain problems—problems that are far more common and pervasive than those posed by encounters with strangers. A problem that arises in all relationships is one of conflicting interests—individuals want to act in ways that benefit themselves at the expense of their partner. This is true for every human relationship, including those between parents and children, siblings, romantic partners, friends, and coworkers. In theory (and maybe also in practice), the easiest way for two individuals to resolve a disagreement is to have a fight. The winner gets what he or she wants and the loser, well, loses. Disagreements between two parties can also be solved by negotiation leading to compromise.

The problem with both strategies is that these ways of settling disagreements can be very costly and are not always effective. Fighting can cause significant damage (both physical and psychological) to the parties involved and to their relationship—possibly leading to its dissolution—while negotiation can entail significant costs in time, energy, and cognitive and emotional resources (for example, constant worrying and rumination). Continuous fighting or negotiation also makes relationships unstable and stressful. Mother Nature—who shapes the minds and behavior of living organisms through natural selection—always tries to find cost-effective solutions to the problems these organisms encounter in their environments. Humans and many other primates live in complex societies in which even the closest and strongest social relationships feature strong competitive elements. Individuals with close social relationships interact regularly, and their interests clash multiple times a day. Yet I know of no primate species in which individuals fight or negotiate all the time. No, Mother Nature has found a better solution to the problem. It's called dominance.

Two individuals in a relationship establish dominance with each other so that every time a potential disagreement arises, there is no need for fighting or negotiation. The outcome is always known in advance because it's always the same: the dominant individual gets what he or she wants and the subordinate doesn't. There is no risk of injury, no waste of time or energy or cognitive or emotional resources. The relationship is stable and predictable, which is good for mental health. The resolution of disagreements through dominance has a cost, of course, but as we'll see later, this cost is paid entirely by the subordinate. If the dominant individual in the relationship is smart, however, he or she will help reduce the cost of losing by making sure the subordinate gets something out of it—or by giving the appearance that this is the case. Before we go deeper into a discussion of dominance, let me clarify that dominance is not the only mechanism for imposing one's viewpoint and interests upon another individual. Other mechanisms for social control, such as coercion or blackmail, are available, but that's material for another story.

Dominance Relationships in Humans and Other Animals

Dominance is an integral part of all human social relationships, beginning a few years after we are born. Infants don't need dominance to control their parents; they cry and scream their heads off, and their parents will do anything to make the noise stop. The moment children begin to understand language, however, parents take advantage of the situation to establish dominance and tell their children what to do. They start giving orders, and when their children cry, parents tell them to shut up. Children remain subordinate to their parents for a few years, as is in their best interest—they are entirely dependent on their parents during this time and wouldn't do well without their support. Children, of course, don't particularly like being subordinate to their parents, but they lack the social skills—and usually the guts—to mount an effective rebellion. They sometimes try to use psychological warfare tactics that were effective at a younger age, like crying and fussing a lot, but older children have lost the leverage that made these tactics successful in the first place, such as blackmail—the threat to harm oneself.[2] Too much crying can be harmful to a baby, whereas the crying of an eight-year-old can be safely ignored, at least for a while.

In the parent-child relationship, the real challenge to parental dominance begins in adolescence. By then children have realized that while baby tactics no longer work, they can do battle on the parents' own turf. This is also the time when evolutionarily it's in the children's best interest to fight for their independence—challenging parents is adaptive for adolescents—so when they push they have Mother Nature backing them up. The struggle for dominance during adolescence doesn't happen the same way in every parent-child relationship. Some parents make concessions to their children and become less authoritarian, but maintain their dominance over their children for the rest of their lives. Some children are successful in reverting the dominance relationship and start calling the shots; their parents acquiesce and accept their new subordinate role. Finally, in some cases, neither party wants to give in, dominance remains unresolved, and parents and children bicker for the rest of their lives—or simply stop speaking.

I mounted a serious challenge to my mother's dominance when I moved out of my parents' house to go to college. I am now forty-seven years old, but my seventy-eight-year-old mother still tries to tell me what clothes I should wear, what food I should eat, and where and when I should go on vacation. Clearly, these are trivial issues, and we don't really fight about them. We don't really fight over this or that: *we fight over who should make decisions over this or that.* That is, we fight over dominance. This happens despite the fact that we now live on different continents and see each other only once a year. (But we do exchange emails.)

Dominance between parents and children is by no means a uniquely human phenomenon. In all animal species in which offspring maintain long-lasting ties with their parents—typically their mothers—these relationships have a strong dominance component. This is true for social insects such as ants and bees, where the queen mother dominates her daughters, as well as many vertebrates, including, of course, other primates. Rhesus macaques live for twenty to thirty years, and daughters spend most of their lives in close proximity to their mothers. As a rule, the mothers are dominant and the daughters subordinate, which makes their relationships very stable. However, when mothers become old and weak, some daughters decide that they've had enough of their mothers' dominance, beat them up, and reverse the dominance relationship once and for all.[3]

Dominance is also a prominent aspect of relationships between siblings. Sibling conflict is widespread in the animal world, as described by zoologist Douglas Mock in his book *More Than Kin and Less Than Kind: The Evolution of Family Conflict*.[4] In species as distantly related as pelicans and spotted hyenas, sibling conflict takes the extreme form of siblicide. Pelican mothers lay two eggs but can afford to raise only one chick, so as soon as the eggs hatch the chicks start fighting, with the result that one ends up killing the other. The same behavior is found in spotted hyenas—mothers give birth to twins, and one cub kills the other (or tries to) within hours or days of being born.

In other species, sibling conflict is not resolved through murder but through dominance. In bird species, parents bring food back to the nest and deposit it directly into the mouths of their begging chicks. Dominance between chicks may make the difference between receiving and not receiving food from the parent. For example, the three eggs laid by female egrets hatch asynchronously, so that the chick born from the egg that hatched first is always larger than the chicks born from the eggs that hatch later. The firstborn chick wins all the fights with its siblings, and when Mom and Dad return to the nest ready to regurgitate their food, the firstborn chick receives the bulk of it. The other two chicks have to fight among themselves for leftovers. Similarly, in mammals that produce offspring in litters, dominance between siblings determines how much milk they obtain from their mother. Piglets start fighting for dominance within a few hours after birth. Those that are born larger and stronger become dominant and occupy the anterior teats, which provide the most milk. The smaller piglets are forced to suck from the posterior teats, which produce less milk. These teat positions are maintained until weaning, with the result that the dominant piglets become larger and larger, and the subordinate piglets remain small. As in other walks of life, the rich get richer, and the poor. . . .

In humans, dominance between siblings is not limited to twins who compete for their mother's milk. All siblings compete for their parents' attention, as well as other resources, and establishing dominance is a way to settle potential conflicts before they arise. In a pair of siblings, dominance may depend on differences in age, gender, or parental favoritism. In general, older siblings are dominant over younger ones by virtue of greater size, strength, and superior social skills. Relationships between preadolescent siblings who are close in age, however, are often characterized by a lot of fighting and dominance instability. Preadolescence is a time when competition for a parent's attention is particularly intense, and it's also the time when younger siblings can challenge their elders with some chance of success. Older siblings sense this, and much of the fighting between siblings involves aggression from the top down to maintain the dominance status quo. The most stable and long-lasting sibling relationships are those in which dominance is clearly established from the beginning and is never challenged—for instance, when one sibling is significantly older than the other. A friend of mine has a brother who is ten years younger than she is. They've always had a stable and close relationship, with little or no fighting. My friend has never felt threatened by her brother and therefore felt no need to do anything to maintain her dominance. Rather, as a result of how clear and stable their dominance relationship is, my friend has taken up a supportive and protective role toward her brother. They phone each other ten times a day!

Dominance between friends can be subtle or very obvious. In children it tends toward the latter. Children begin competing for status and trying to establish dominance over their peers as early as two years of age.[5] Dominance between children is important because it may determine who gets the attention of an adult, a preferred partner, or access to toys and other valued resources. As soon as children make their first friend, they fight hard to become dominant in the relationship. (Children, especially boys, often use physical aggression to establish dominance.) In my elementary school years, I had a best friend, Massimo, with whom I played after school almost every day. We obviously liked each other a lot and enjoyed playing together, but we were also very competitive. One thing we regularly competed for was the attention of a third boy named Valerio. Every day Massimo and I wrestled like wildcats on the floor of my bedroom and tried to force each other to admit defeat. In Valerio's presence, we would also tease and denigrate one another and attempt to play with Valerio while keeping each other out of the game.

Such indirect competitive tactics are more commonly used by young girls, who don't use direct confrontation and physical aggression as much as boys do. Girls dominate a potential rival by spreading nasty rumors aimed at damaging her reputation; by excluding, ignoring, and socially isolating her, thus making her an unattractive social partner to other girls or boys; or by actively disrupting her attempts at alliance formation. Dominant children, both male and female, use both aggression and affiliation strategies to establish and maintain dominance: they simultaneously attack the child they want to dominate and befriend others who might help them in the process, forming alliances with them. These Machiavellian strategies are similar to those used by other primates to achieve and maintain dominance over their group members. Once Mother Nature has found something that works well in one species, she is happy to use that same trick in other organisms as well.

Dominance in romantic or married couples is an important but underappreciated phenomenon. The most stable romantic relationships and marriages seem to be those in which dominance is clear from the beginning. The dominant partner makes all the decisions, from what show to watch on TV in the evening to where to go on vacation in the summer, and the subordinate partner acquiesces and takes a supporting role. If what people expect from marriage is not necessarily everlasting passionate love but a stable partnership that will allow joint ventures such as buying a home and raising children together, or an opportunity to concentrate on one's career without worrying about house chores, then an asymmetrical relationship with uncontested dominance probably guarantees the best outcome. The secret to a stable marriage is that one of the two spouses must be willing to pay a disproportionate share of the price for the stability.

One problem with such an unbalanced relationship is that once the children are out of the house, career goals have been accomplished, and the mortgage on the mansion has been paid off, the stable relationship may no longer have a reason to exist. The dominant spouse, or both, may lose interest and begin looking for another partner. Another potential problem is that the dominant spouse may become dictatorial and abusive. The subordinate partner can benefit enough from the stability and support obtained from the relationship (and the accomplishment of other goals that comes with it) to compensate for the lack of decisional power and all the losses associated with it, but only as long as the dominant partner adopts a tolerant and respectful dominance style. Abusive dominance makes the costs of subordination skyrocket to the point that the benefits of the relationship are no longer worth it and the subordinate must walk out.

"The course of true love never did run smooth," says Lysander in Shakespeare's *A Midsummer Night's Dream*. Individuals who fall in love with each other and are looking for more than a business partnership face formidable challenges, particularly if they both have strong personalities. If neither individual is willing to take a subordinate role, every time conflicting interests arise or decisions need to be made, the relationship is potentially threatened. In the absence of a clear-cut dominance relationship to settle all the contests, the costs of continuous fighting or negotiation inevitably take their toll. It is common knowledge that when couples fight and eventually break up, they do so over seemingly trivial issues. Clearly, it's not disagreements over the dinner menu or the remote control that lead a couple to divorce. It's the disagreements over who is in charge and who isn't, and the stress and disruption that come along with these disagreements. Couples with unresolved dominance may last for a while, maybe even forever, but their relationship is inherently unstable.

Studies of baboons conducted by Stanford University biologist Robert Sapolsky illustrate the costs of unstable dominance relationships. Savanna baboons live in a complex and competitive society in which success depends on a good dose of selfishness coupled with the ability to form political alliances with others. Adult male baboons have cooperative and competitive relationships with each other, just as Washington politicians do. A study by University of Minnesota zoologist Craig Packer in the 1970s showed that male baboons often form coalitions to create the opportunity to mate with a female. If a female is in estrus and closely guarded by a dominant male who won't allow anyone else near her, two other males may work together so that while one of them picks a fight with the dominant male, the other takes advantage of the distraction to mate with the desired female. The next day the lucky male may reciprocate the favor to his partner. Although male baboons need each other for these political alliances, they also compete for dominance. Thus, even between the closest of allies, the subordinate male is always on the lookout for the right opportunity to challenge the dominance of his buddy. Sapolsky showed that when a dominance relationship is stable, the dominant male has lower levels of the stress hormone cortisol in his blood than the subordinate. When dominance is unstable, however, both individuals have equally high cortisol levels. Males who are being successfully challenged for dominance by their subordinates have the highest levels of cortisol in their entire group.[6]

A reversal of a dominance relationship can have life-changing consequences, for baboons as well as for people. In the 1935 novel *Auto-Da-Fé*—a masterpiece of European literature for which its author, Elias Canetti, won the Nobel Prize in 1981—the main character is a reclusive scholar, Peter Kien, who spends all his time in his apartment, where he keeps a massive library with thousands of volumes.[7] The only person he keeps around is his housekeeper, an illiterate older woman named Therese who rents a room in his apartment and cleans and cooks for him. For eight years, their relationship is straightforward. Kien is the employer and Therese is the employee, he owns the apartment and she is the guest, he is a scholar and she is illiterate. He is dominant and she is subordinate: he barely looks at her when they talk, and she treats him with great deference. However, when Kien misinterprets Therese's conscientiousness in dusting his books for a love of knowledge similar to his own and decides to marry her, their relationship shifts dramatically: they are no longer employer and employee, but husband and wife. All hell breaks loose. Therese is no longer intimidated by Kien's intellectual superiority and believes that her cravings for expensive new furniture and clothes should take precedence over his desire to acquire more books and knowledge. She becomes more and more confrontational with him until one day she loses it and beats him to a pulp. Now their dominance relationship is reversed. Kien is afraid of Therese and becomes passive for fear of another beating. Therese gains control of the apartment and buys all the furniture she wants with Kien's money. When the money runs out, she kicks Kien out of the apartment and pawns his books. If you like novels with a happy ending, *Auto-Da-Fé* may not be the book for you. Canetti is not optimistic about human beings' ability to communicate with one another and resolve their disputes amicably; instead, he painstakingly describes how our lives crumble to pieces when we fall prey to the dark survival instincts of our own minds or those of the people around us. Although Canetti doesn't mention the word *dominance* once in his book, he provides a vivid illustration of the powerful influence that a change in a dominance relationship can have on people's lives.

Dominance Relationships 24/7

Dominance is so intrinsic to human social relationships that we don't even notice it. However, I suspect that if I asked you to make a list of one hundred people you know, including family members, friends, and coworkers, and indicate whether you are dominant or subordinate in your relationship with each of these people, you could give a clear answer for at least ninety-five of them. Normally, we don't think about how our daily interactions with the people we know are affected by our being dominant or subordinate. The truth is, however, that dominance permeates many aspects of our everyday social lives. The same is true for other primates as well.[8]

In a group of savanna baboons, every individual has social relationships—and dominance relationships—with everybody else in the group. In baboons, one practical consequence of dominance is "priority of access": if a dominant and a subordinate want the same thing—a piece of food, an attractive mate, or a spot in the shade on a hot summer day—the dominant will always get it, or get it first. The dominant baboon may let the subordinate have something they both want only if he or she cares much less about it than the subordinate does—for example, a piece of food when the dominant's

31

belly is already full. Although the importance of dominance is particularly clear when two individuals want the same thing or disagree about something, a dominance relationship between two individuals operates twenty-four hours a day, seven days a week, and affects how the two individuals interact with each other in virtually every situation.

To begin with, if we measure the amount of time two male baboons spend looking at each other on a given day, chances are the subordinate spends a lot more time looking at the dominant than vice versa. Moreover, the subordinate is more likely to change his behavior in response to the dominant than the other way around. For example, if the subordinate is sitting in a corner eating a banana and the dominant walks by, the subordinate is likely to stop eating and go sit in a different spot. If the dominant is eating a banana and the subordinate walks by, the dominant will probably continue eating the banana as if nothing has happened. More generally, the subordinate avoids the dominant and gets out of his way, while the dominant pays no attention to the subordinate. When dominant and subordinate cross paths, the subordinate is likely to greet the dominant with a "bared-teeth display" or by "presenting" his behind. The dominant rarely greets the subordinate. The dominant stares at the subordinate or uses other threatening facial expressions or vocalizations and may even attack the subordinate. The subordinate never initiates threats or aggression toward the dominant, although when under attack he may occasionally fight back in self-defense.

In primate societies in which individuals have strong and stable dominance relationships, most fighting consists of intimidatory aggression by dominants against subordinates to maintain and reinforce the status quo. To try to control a dominant's aggression, win his tolerance, and possibly receive some favors, a subordinate not only behaves submissively but also provides services to the dominant. For baboons and other primates, these services consist mainly of grooming. The subordinate is willing to spend hours cleaning the fur and massaging the body of a dominant. Generally, what the subordinate receives, or hopes to receive, in exchange is more grooming, tolerance, or help. The dominant allows the subordinate to stay close to him during or after grooming, and if the subordinate gets into a fight with another individual and calls for help, the dominant may intervene and lend a hand. In some cases, there is an exchange of grooming between a dominant and a subordinate, but, as described earlier in this chapter, the grooming given and received between the two is never well balanced.

Many of the baboon behaviors I describe here have obvious parallels in humans. I recently witnessed a conversation at a coffee shop on campus between two female colleagues I know very well: one is a tenured professor in her sixties—let's call her Jane—and the other a young and untenured assistant professor hired a few years ago—I'll call her Jill. When they found a table where they could sit, I noticed that they both went for the chair with its back against the wall—people like to have their back protected when they sit in a coffee shop or a restaurant—but then Jill immediately withdrew and let Jane take the favored chair. During the conversation, Jill was very attentive to everything Jane said and did, maintaining almost continuous eye contact with her, while Jane's attention wandered when Jill was talking. Jill also smiled at Jane more frequently than the reverse. At some point, I overheard them talking about a potentially contentious issue—the hiring of a new faculty member in their department. As Jane forcefully stated her opinion on the subject to Jill, she stared her down and raised her tone of voice. Jill smiled harder than ever, quickly deferred to Jane's viewpoint, and immediately moved the conversation to a more neutral topic. She also offered to get some more milk for Jane's coffee, and toward the end of the conversation she apologized profusely for not being able to stay longer. Before getting up from her chair, she waited until Jane stood up first, and then they both left the coffee shop, Jane walking out the door first, and Jill following behind.

I once had a male colleague who stopped walking every time we crossed paths in the hallway of our building and flattened himself with his back against the wall. He acted this way with other people too. He never walked in the middle of a hallway; rather, he was always way over to one side, with his back sliding against the wall. I once observed a male pigtail macaque who walked around the same way, his back always flattened against the wall of his compound. Like my colleague, this male was subordinate in the dominance relationships he had with all the other adult males within his group.

Pecking Orders

In rhesus macaques, baboons, and other primates, dominance generally follows the transitive property: if A is dominant over B and B is dominant over C, then A is also dominant over C. As a result, all individuals can be ranked in a *linear* dominance hierarchy, with the individual who is dominant over all the others at the top and the individual who is subordinate to all the others at the bottom. An individual's position in the hierarchy is called its *dominance rank*. To emphasize the distinction between dominance in a dyadic (pair) relationship and dominance rank in a hierarchy, students of primate behavior use the terms *dominant* and *subordinate* to refer to roles in a relationship and the terms *high-ranking* and *low-ranking* to refer to an individual's position in the hierarchy.

Dominance hierarchies need not necessarily be linear. If a dominance relationship is nontransitive, such that A is dominant over B and B is dominant over C, but C is dominant over A, then the dominance hierarchy is *nonlinear*: it contains triangles or loops involving three or more individuals. Finally, there can also be *despotic* hierarchies, in which one individual rules over all other members of the group, with no rank distinctions being made among others.

Researchers interested in understanding the consequences of dominance have examined and compared the lives of high-ranking and low-ranking monkeys. Although in special situations—for example, when primates are kept in captivity and provided with abundant food and protection from danger—it may look as though high-ranking and low-ranking individuals lead very similar lives, in more naturalistic conditions the advantages of high rank are substantial. High-ranking animals survive longer, reproduce better, and generally live a healthier, more comfortable, and less stressful life than the low-ranking ones. It's the same for humans. For an entertaining description of the lives of the powerful and the powerless in human societies, as well as the effects of dominance hierarchies in the workplace, I highly recommend Richard Conniff's books *The Natural History of the Rich* and *The Ape in the Corner Office*.[9]

Dominance and hierarchies are not unique to primates. According to Harvard evolutionary biologist Edward O. Wilson, dominance in animals was first discovered by Swiss and Austrian entomologists who were studying bumblebees in the 1800s.[10] These researchers reported that the queen was dominant over the worker bees and that subordinate workers that tried to steal and eat eggs were physically punished by the queen or other more dominant individuals. Insects aside, the first well-described dominance hierarchy in animals was the "pecking order" in chickens studied by the Norwegian biologist Thorleif Schjelderup-Ebbe between 1920 and 1935. He showed that when a bunch of chickens are first thrown together, they fight with each other over access to food, nest sites, and roosting places. When fights between two individuals have a clear winner and loser, aggression between them ceases and in subsequent days the loser always yields to the winner. Schjelderup-Ebbe demonstrated that chickens could recognize each other and remember the outcome of fights with particular individuals for weeks. He also described how the dominant chickens maintained their status by either pecking an opponent or making a threatening movement toward an opponent with the intention of pecking. Shortly after his work was published, numerous other researchers documented the occurrence of dominance relationships and hierarchies in other birds and mammals.

Confusion About Primate Dominance

Dominance in primates is nothing special compared to other animals, but primatologists have differing opinions about it. In the early days of primatology, one group of researchers believed that dominance was simply the result of some individuals acting tough, threatening others and picking fights, and inducing fear in less aggressive individuals. Others held the opposite view: the fearful and submissive behavior of subordinates accounts for dominance relationships; aside from being aggressive, the dominants don't contribute much. The idea that dominance can simply be equated with the behavior of dominants or the behavior of subordinates was later criticized by primatologists who argued that monkeys and apes have social relationships, just like people, and that dominance should be considered a property of relationships, not of individuals. This means that an individual can be dominant in a relationship with one individual and subordinate in a relationship with someone else.

But not everyone agreed. In a short paper published in 1981, primatologist Stuart Altmann, now an emeritus professor at Princeton University, argued that monkeys and apes don't have dominance relationships with one another simply because they don't have social relationships. According to Altmann, social relationships are abstractions that exist only in the minds of the people who study primate behavior—an invention of the human observer to describe why monkeys behave the way they do. Monkeys don't categorize individuals as dominants or subordinates, high-ranking or low-ranking, kin or nonkin, friend or foe, Altmann wrote. Any individual's behavior is always a response to another individual's behavior. Since dominance relationships don't exist, they cannot have any influence on the behavior, survival, or reproduction of individuals. He concluded that dominance relationships are important, but only to the researchers, not to their subjects. Altmann applied his reasoning to nonhuman primate behavior, but psychologists known as *behaviorists* have similar views about human behavior as well. From their perspective, an unhappy husband and wife respond to each other's annoying behavior, but their relationship is an abstraction that exists only in the mind of their marriage counselor.

Paradoxically, the extreme view that dominance isn't real but exists only in the minds of researchers was also held by Robert Hinde, a prominent British ethologist and now an emeritus professor at the University of Cambridge, who in the 1970s convinced many other primatologists that monkeys and apes do have social relationships.

Figure 2.2. *Dominance and anxiety as intervening variables.*

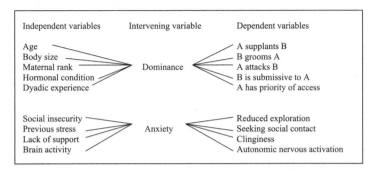

While he was writing on various aspects of primate social relation-ships, Hinde also published a series of articles in which he argued that dominance "does not exist in a concrete empirical sense but it may have usefulness as an explanatory concept." In his view, domi-nance should be considered an "intervening variable" that makes it easier for human researchers to explain animal behavior. As illustrated in Figure 2.2, when individuals A and B enter a contest over a re-source, their relative age, their body size, the social status of their mothers, their hormone levels, and their previous interactions (the independent variables) can lead to an interaction in which A sup-plants B, B grooms A, A attacks B, B is submissive to A, or A gets what he wants (the dependent variables). Hinde argued that, when researchers attempt to explain behavior, instead of analyzing the twenty-five different lines that connect the five independent variables to the five dependent ones, it's easier to place an intervening variable in the middle, in this case dominance. Hinde believed that researchers make up a lot of intervening variables to explain behavior. According to him, emotions don't exist in a concrete sense either. For example, as illustrated in Figure 2.2, anxiety is an intervening variable we can use to explain why lack of self-confidence and the absence of a sup-porting adult can lead a child to spend less time exploring and to seek social contact with other children in novel environments.

Most primatologists these days disagree with Altmann and Hinde and accept that dominance is real; pretty much everyone agrees that anxiety is real too. Why else would so many psychiatrists prescribe Xanax to their patients? There is still disagreement, however, as to what dominance relationships really are. For example, according to primatologist Irwin Bernstein, dominance is a *learned* relationship. In this view, dominance relationships are established in the course of the initial encounters between two individuals by means of repeated fighting. Once fights have a clear winner and loser and dominance is established, signals are exchanged to inform others that previous events are remembered and the dominance relationship is acknowledged and respected. As we'll see later, although previous experience is generally an important component of dominance, there are situations in which experience and learning don't play a role in the establishment of dominance.

Primatologists recognized early on that—as explained earlier—dominance between two individuals can be expressed in different ways: priority of access, directionality of aggressive and submissive behavior, freedom of movement and spatial position, lopsided exchange of grooming, visual monitoring, and receipt of attention. Instead of simply recognizing that dominance is a phenomenon with multiple dimensions, some primatologists have argued that there are different types of dominance. For example, Frans de Waal has suggested that primates have two kinds of dominance relationships: *real dominance*, which explains who wins fights, and *formal dominance*, which explains why subordinates give submissive signals to dominants. Wilson and others have proposed that there are different types of dominance relationships depending on context. One type of dominance, referred to as *absolute dominance*, is independent from context. If individual A has absolute dominance over individual B, A always gets priority regardless of whether A and B are fighting over food, space, or a mate and regardless of where the fight takes place. *Relative dominance*, by contrast, is dependent on context. For example, A could be dominant over B when they both want the same food, whereas B could be dominant over A when they both want to mate with the same attractive female. One particular type of relative dominance is *territoriality*, where dominance depends on the location of the encounter. In animals that defend territories in which they build nests and forage, A may be dominant over B when their encounter takes place in A's territory, and B is dominant over A when the encounter takes place in B's territory. Another distinction was recently introduced by primatologist Rebecca Lewis, who proposed that a dominance relationship between two individuals should actually be called a *power relationship* and that we should distinguish between *dominance*, when power is based on force or the threat of force, and *leverage*, when power is based on resources that cannot be taken by force.[11]

What should we make of all these distinctions between different kinds of dominance? In my view, they are not justified and are a potential source of confusion and misunderstanding. The best tool to clear the muddy waters of primate dominance is game theory. Once we consider dominance using game theory, it becomes clear that not only are dominance relationships real, but dominance is always one and the same thing.

Of Bears, Hawks, and Doves

The use of game theory to examine dominance was pioneered by British evolutionary biologist John Maynard Smith in the early 1970s with the publication of his classic article, "The Logic of Animal Conflict."[12] There he looked at how two individuals decide whether to settle a disagreement by having a fight or by establishing dominance. Before I go on, let me clarify that even though evolutionary biologists use terms such as "decisions" and "logic," they don't presume any conscious rational thinking on the part of animals (or of people, for that matter). The decisions and the logic in question are the products of natural selection, which has given organisms predispositions to behave adaptively (in a way that increases the benefits and reduces the costs of their behavior) without necessarily engaging in complex thinking or being aware of the consequences of their actions.

To illustrate the game theory approach to dominance, let's start with a simple hypothetical situation. Yogi Bear and Boo-Boo Bear (the famous cartoon characters created by Hanna-Barbera) meet in a forest one day and there discover an apple. They have never met before and have no reason to think they will meet again in the future. They both want the apple, but the apple cannot be shared. For the sake of simplicity, let's assume that if Yogi and Boo-Boo fight over the apple, they have exactly the same probability of winning. The moment they see each other and the apple, each bear has the same two options: to fight or to let the other bear take the apple. If a fight is initiated, the winner will eat the apple and the loser will get nothing. If one bear yields, he will simply walk away, maybe to look for an apple somewhere else. If both bears yield, they will use peaceful signals, such as facial expressions or gestures, to negotiate who gets the apple and who doesn't. In this situation, each bear has the same probability of getting the apple so that, were they to meet one hundred times, each bear would get the apple about fifty times.

At this point, some game theory terminology will make it easier to discuss this scenario and apply it more broadly to other situations. An apple is a *commodity* that two individuals may want at the same time. In the language of game theory, the options to escalate the fight or to yield are called the *Hawk* and *Dove strategies*. A contest in which the two opponents have the same probability of winning is called a *symmetrical contest*.

Whether Yogi and Boo-Boo play the Hawk or the Dove strategy depends on the value of the apple (let's call it Benefit, or B) and the cost of fighting (Cost, or C). The benefits are the intrinsic nutritional value of the apple and the potential of the apple to keep a bear from starving. (The same apple with the same caloric content is a lot more valuable to a bear that is about to die from starvation than to a bear that has just eaten ten pounds' worth of honey.) The main cost of fighting is the risk of injury, even death. Fighting also has other secondary costs: one can burn a lot of energy, make noise that attracts enemies or predators, or miss a chance to look for other food or a mate. In the case of two individuals who are friends (although this doesn't apply to Yogi and Boo-Boo), one possible cost is damage to the friendship and the potential loss of future benefits it may have

Figure 2.3. *Yogi Bear and Boo-Boo Bear.*
Cartoon by Matthew Hunter.

brought. Game theory predicts that when the benefit of eating the apple is greater than the cost of fighting (which is the same as saying that their ratio is greater than 1, or B/C > 1), both bears will play Hawk and escalate the fight.

When the cost is greater than the benefit (B/C < 1), the situation is more complicated. Let's assume that fifty pairs of bears meet at the same time and get into a potential conflict about fifty different apples. Some bears are expected to play Hawk and others to play Dove. It can be shown mathematically that in this imaginary population of one hundred bears playing the Hawk-Dove Game, the frequency of Hawks is equal to the ratio between Benefit and Cost (B/C), and the frequency of Doves is equal to 1 − B/C. For example, when B/C is 0.60, sixty out of one hundred bears will play Hawk and forty of them will play Dove. When the benefits are almost as high as the costs and B/C is very close to 1, most bears in the population will be Hawks and only a few will be Doves. When the benefits are negligible compared to the costs and B/C is close to 0, almost all bears will be Doves.

The Hawk-Dove Game is a simple model, but real life is never so simple. In real life, a truly symmetrical contest doesn't exist. Instead, there are always differences, or *asymmetries*, between the two contestants so that one of them has a greater probability of winning a fight than the other. In practice, this means that when Yogi and Boo-Boo confront each other over the apple, both immediately recognize that if they have a fight, the probability of winning or losing would not be fifty-fifty. The bear who thinks he has a higher chance of winning is more likely to play Hawk and provoke a fight—for example, by threatening his opponent—while the bear who thinks he has a higher chance of losing is more likely to play Dove and give up the apple peacefully. The moment the two bears recognize that their chances of winning and losing are not the same and act accordingly, their dominance relationship is established: one individual is dominant and the other is subordinate.

Establishment of dominance can occur when two individuals meet for the first time—as soon as they look at each other—or it may take place after repeated encounters and some fighting. Sometimes it's not immediately clear to the two individuals whether their chances of winning or losing the fight are the same; an empirical test is needed. If Yogi Bear wins ten consecutive fights against Boo-Boo Bear, when they meet for the eleventh time they may both recognize that Yogi has a better chance of winning the next fight than Boo-Boo does. At this point, Boo-Boo yields to Yogi, and Yogi is happy to eat the apple without beating Boo-Boo. When dominance is established, the fighting usually ends. But in some cases, the two bears continue to think that their chances of winning are the same, or each bear thinks that he has the upper hand. In this case, dominance is not established and the bears keep on fighting.

It's important to understand that establishing dominance is advantageous to both contestants and always preferable to fighting. Resolving the dispute with dominance is good for Yogi, the dominant bear, because he gets the apple without paying the price of fighting. Resolving the dispute with dominance is also good for Boo-Boo, the subordinate bear, although not as good as it is for Yogi. By yielding to Yogi, Boo-

Boo doesn't get the apple, so he gets zero benefits. However, if Boo-Boo Bear had fought and lost, not only would he have lost the apple but he would also have paid a high price physically. By becoming subordinate, Boo-Boo reduces the costs he is likely to pay. So the advantage of establishing dominance for the subordinate is that he cuts his losses.

Cutting one's losses? That's it? Yes, the truth is that subordination sucks, and I wouldn't recommend it to anyone. There are two ways, however, in which Boo-Boo might further benefit from yielding to Yogi. First, if Yogi is a generous bear and the contest with Boo-Boo is over a commodity that can be shared—say, an apple pie—he may let Boo-Boo eat a small piece of the prize or give Boo-Boo something else as a token of appreciation for conceding. The other advantage of subordination is that by yielding to Yogi that day and other times in the future, Boo-Boo saves ammunition. He can safely wait things out until one day he is able to successfully challenge Yogi and reverse their dominance relationship. In fact, it would be very bad if Boo-Boo didn't even try once to challenge Yogi. Patience is a virtue for a subordinate, but resignation may be the kiss of death. To understand the factors that might lead to a dominance reversal, we must delve further into these asymmetries that determine dominance.

There are two kinds of asymmetries between two individuals engaged in a contest. One type of asymmetry has to do with individual attributes, such as physical characteristics. Yogi is taller, heavier, stronger, healthier, and more mature than Boo-Boo. Characteristics that make an individual more likely to win or lose a fight are called *resource holding potential*, or RHP. Two individuals can recognize differences in their RHP as soon as they meet for the first time. When Yogi and Boo-Boo first met, both immediately noticed that Yogi was twice as big as Boo-Boo. Hence, they immediately established dominance. Yogi played Hawk and threatened to fight; Boo-Boo played Dove and acted submissively. Thus, a history of previous interactions is not necessary to establish dominance, and dominance is not necessarily a learned relationship.

In the experiment I described in Chapter 1, I mentioned that in the pairs of monkeys that had never met before the two females appeared to be sizing each other up. In some of these pairs, one of the two monkeys "smiled" submissively to the other and started grooming her. The two monkeys probably recognized that there were asymmetries in RHP, perhaps in size, and the individual who perceived herself to be inferior showed a sign of fear and submission, the bared-teeth display. The immediate establishment of dominance allowed the grooming exchange to begin. As expected, the exchange of grooming was heavily lopsided, with the subordinate monkey doing most of the work and the dominant failing to reciprocate. In other unfamiliar pairs, no clear asymmetry in RHP was communicated or recognized. As a result, dominance was not established and little or no grooming was exchanged. Both monkeys "defected" and looked stressed out when their hour together in the cage was finally over, presumably the price they paid for their inability to establish dominance—just like two spouses who are both unwilling to give in and pay the price for the stability and peace of their marriage.

The outcome of a fight between two individuals can be determined not only by differences in size and physical strength but also by differences in their *willingness* to use their bodies and take risks. Differences in motivation may arise from differences in the value of the commodity being contested: if the commodity is more valuable to one opponent, then this individual will be willing to fight harder than the other. In game theory language, an asymmetry in the value of the commodity to the two opponents is called an *asymmetry in payoffs*; the individual who has more to gain from winning is also more motivated to fight.

Most, if not all, kinds of context-dependent dominance reflect asymmetries in payoffs. In territorial species, territory holders or residents are more motivated to fight against intruders who have invaded the territory because they have more at stake. Their territory contains their nest, their food, and often their mates and offspring. Losing one's territory may mean losing everything. The intruder has a lot less to lose; if things don't work out here, he'll simply invade the territory next door. Once again, decisions to fight or yield don't require mental calculations about the costs and benefits of particular moves.

Figure 2.4. *Bared-teeth display shown by a crested black macaque,* Macaca nigra. *Photo courtesy of Dr. Irwin Bernstein.*

For example, residents' threatening to fight or escalating a fight against an intruder may simply be an emotional response to the situation. The sight of an intruder trespassing into his territory may enrage the resident, and this emotion enhances his aggressive motivation. Conversely, being in an unfamiliar place induces fear in the intruder and increases the probability of submission.

By this point it should be evident that there aren't multiple kinds of dominance after all. Dominance is always one and the same, but dominance interactions may be different owing to asymmetries in RHP or in payoffs between contestants. Game theorists have developed more complex versions of the Hawk-Dove Game to take these asymmetries into account. As with the simple Hawk-Dove Game, interaction outcomes are predicted by the value of the commodity and the cost of fighting. Clearly, the usefulness of these models to explain the behavior of individuals in real-life situations depends on how accurately one can measure the exact value of B and C to the two opponents. Multiple asymmetries between the two opponents can make the measurement of B and C very tricky. Moreover, there is a further complication. For the sake of simplicity, the game theory models of asymmetric contests assume that two opponents always have accurate information about their asymmetries and relative probabilities of winning or losing a fight. Again, real life is hardly so simple, and the information in question is not always accurate.

Theatrical Displays

When Yogi Bear and Boo-Boo Bear meet and enter a contest over the apple, they don't just stare each other down and size each other up to gauge physical strength and level of anger. Chances are that each bear will exhibit behaviors to affect the other's assessment. Logic says that if asymmetries lead to the establishment of dominance and this is advantageous to both contestants, then all individuals should have an interest in making sure that the asymmetries are effectively communicated and understood. So when two bears want the same apple, it would be advantageous to both parties if each honestly communicates any asymmetries to the other: "I'm strong"/"I'm weak"; "I'm starving"/"I've just had a big dinner"; or "I'm willing to fight to the death"/"I have a headache and I'm in no mood for fighting." Given that communication of RHP and motivation through behavior is advantageous, natural selection has favored the evolution of particular behavioral signals that have this function: they are called *behavioral displays*, to distinguish them from physical displays of RHP such as large antlers in deer or large canine teeth in male lions. Animals have evolved facial expressions that communicate emotions such as anger or fear and give a potential opponent some information about their motivation to fight or to retreat. Animals (and some people) also scream or break things as a demonstration of strength.

However, this all assumes candid communication. If communications about asymmetries through behavioral displays were always honest, dominance interactions would be predictable with the application of Hawk-Dove models for asymmetric contests. We would all live in a far less stressful world in which parents never squabble with their children and celebrity couples never make the cover of tabloid magazines with their fights. The problem is that communication of asymmetries isn't always honest. Since the benefits of being dominant are greater than the benefits of being subordinate, it is advantageous to cheat and communicate exaggerated, or blatantly false, information about one's RHP and willingness to fight. Thus, natural selection has rewarded the tendency to bluff, and bluffing has become an important determinant of dominance. Natural selection has also rewarded the ability to be skeptical and detect bluffing because skeptical individuals have more chances of becoming dominant than gullible ones.

When two animals encounter each other for the first time, a fight will ensue if no clear asymmetries are communicated, recognized, or believed. Winning or losing a fight is a clear demonstration that asymmetries exist. For pairs of individuals that need one or multiple fights to establish dominance, dominance may become a learned relationship. Once dominance is established, these individuals exchange behavioral displays with each other whenever they meet to refresh their memories and to communicate that nothing has changed: the asymmetries of the past persist in the present. Dominants also use periodic aggression to refresh the subordinates' memory. Using the language of learning theorists, if subordination is a learned response, periodic reinforcement is necessary to prevent the extinction of the response. Depending on the extent of the asymmetries and the kind of social system in which individuals live, the maintenance of dominance may be driven mostly by dominants, or by subordinates, or by both. When asymmetries between opponents are small, maintenance of dominance may be mostly driven by dominants with frequent threats or aggression. When asymmetries are large, subordinates probably live in a constant state of fear, and it's in their own interest to express frequent unsolicited subordination to dominants. Even in the most asymmetrical dominance relationships, however, rebellion from subordinates is always a possibility, because the asymmetries between two individuals may change over time or in different situations. If an event makes it possible to differentiate asymmetries in the present from asymmetries in the past, fighting may escalate and dominance may be reversed. Subordinates may reach a point where they have nothing left to lose and therefore are more motivated to fight.

Resource holding potential can also be enhanced by acquiring physical strength or political power. In many species of primates, including humans, contests that initially involve two individuals can rapidly escalate to involve the individuals' families and allies. In the matrilineal society of rhesus macaques, for example, dominance relationships between females are established on the basis of the support provided by female relatives. Therefore, the size and power of one's family can be considered part of an individual's RHP and a source of important asymmetries. Asymmetries that arise from the availability of support may need experience to be recognized and taken into consideration. Through experience, a subordinate macaque female learns that higher-ranking females have more numerous and/or more powerful allies who will come to their aid. But like any other asymmetry, the availability of political power must be continually monitored because political allies may unexpectedly turn their backs.

Born Leaders and Losers?

There is another complication. Although it's true that dominance is a property of relationships, not of individuals, it's also true that individuals' physiological and psychological characteristics contribute to their RHP and to their motivation to fight; some may be predisposed to act dominant and others to act subordinate. For example, individuals who are born with low levels of serotonin in their brains are predisposed to be impulsive and aggressive, and having a lot of testosterone in one's body makes people competitive and driven to succeed. Children who show sharp increases in cortisol when something upsets them may be less capable of handling the stress associated with agonistic confrontations later in life—and are therefore more likely to avoid them.

Predispositions to act dominant or subordinate don't involve only emotions and physiology but cognition as well. In humans and some other primates, dominance depends on social and political intelligence. Cognitive skills that lead to dominance include the ability to learn the rules that constrain behavior in a social group; interpreting, predicting, and manipulating the behavior of others; and forming powerful alliances based on reciprocal obligations. Individuals who become dominant learn better and more quickly than others the rules that govern group behavior, whereas those who become subordinate break these rules because in some cases they are not even aware that the rules exist. In humans, dominance also depends on the ability to decode nonverbal behavior (what facial and bodily cues reveal about an individual's emotions and motivation), to guess what goes on in other people's minds (such as what they know and don't know, what they want and don't want, what they believe and don't believe), and to deceive. Dominant individuals are better at interpreting others and persuading them to do what they want with all possible means, including deception. The ability to charm and befriend others and to form alliances with them through the exchange of favors is also crucial. Considering that autism spectrum disorders—which involve deficits in social intelligence skills—are highly heritable, it is likely that individuals at the other end of the continuum, who excel at these skills, may have their genes to thank for some of their talents.

Predispositions, of course, can also be the product of experience, and behaving in a dominant or subordinate fashion can be learned. Being born with a predisposition doesn't necessarily mean that this predisposition comes straight from our genes. There are environmental effects on us during the nine months we spend in our mother's womb—for example, fetuses are exposed to different amounts of testosterone and cortisol in utero, and this exposure may affect their brain development. Sometimes environmental effects are mediated by genetic mechanisms—being exposed to particular environments may lead to the expression of certain genes and the suppression of others. Obviously, after we are born, there are myriad opportunities for environment and experience to affect our propensity to act dominant or subordinate. One particularly important type of experience involves confrontations in themselves, and their outcome.

In the males of many animal species, including monkeys, apes, and humans, winning a fight raises testosterone levels, while losing reduces them. Having high or low testosterone, in turn, increases the probability of winning or losing the next fight, respectively. This means that Boo-Boo Bear is more likely to defeat Yogi Bear if he has just defeated Winnie the Pooh. Yogi is more likely to lose against Boo-Boo if he has already suffered another loss.

The effect of a previous victory or defeat on subsequent confrontations need not be dependent on testosterone. There are other physiological and psychological changes that follow victory or defeat. Agonistic confrontations are stressful for everybody, and as we saw before, in savanna baboons the cortisol levels of all members of a group rise during periods of social instability, when some individuals are challenging others for rank. Cortisol goes back to low levels in dominants once the confrontations are over, but it stays elevated in the subordinates if they continue to be harassed and intimidated. Adults and children of low socioeconomic status (SES) have higher cortisol than those of high SES and display greater changes in cortisol and blood pressure levels in response to a conflict. In married couples, the spouse who perceives himself or herself to be subordinate in the relationship shows a higher increase in blood pressure reactivity during marital disagreements. Just as an increase in testosterone following a victory may increase one's self-confidence, ambition, and motivation to fight again, a decrease in testosterone and an increase in cortisol following repeated defeats or in association with chronic subordination can result in depression and shame, which can lead to submissive behavior: avoidance of eye contact, hunching body posture, social avoidance. These physiological and behavioral changes promote acceptance of and adjustment to subordination.[13]

By now it should be clear that dominance and submission don't exist only in the minds of the researchers who study behavior but in fact are deeply lodged in human minds and human bodies as well as in the minds and bodies of many other animal species. We have physiological, emotional, and learning mechanisms that allow us to constantly assess our own performance in agonistic confrontations, that tell us whether we are dominant or subordinate in a relationship, and that help us adjust to the situation. When we win a contest and become dominant, we feel good about it and we want more of the same. When we lose and become subordinate, we feel bad and either cut our losses and adjust to the situation or start preparing for a future rebellion.

The experience of wins and losses with other individuals influences not just individuals' perception of their own RHP but also their assessment of other individuals' RHP and their estimation of their probability of winning or losing. After a crushing defeat, any new opponent looks scary to us. Or, if the bully who defeated us was a bold man with a gray beard, we may behave submissively to all bold men with gray beards we encounter in the future. Since we project changes in our internal physiological and psychological states onto others through our behavior, our own self-assessment also influences how our RHP and motivation to fight are assessed by others.

As we all know, social relationships can be quite complex. The interplay between individual characteristics, context, and previous experience, with all of their feedback mechanisms, makes the assessment of asymmetries in a contest very difficult. Dominance, however, is an integral component of all our social relationships and has a pervasive influence on all aspects of our everyday social lives. The sooner we recognize this, the better we will understand why our relationships work the way they do. Humans and some other primates are obsessed with dominance, although not necessarily at a conscious level. Dominance is so entrenched in human nature that thinking we can have social relationships without it is unrealistic. What we can do instead is to provide mechanisms that allow all individuals to fully express their own potential—whether in their personal or public life—so that they can compete for dominance to the best of their abilities. Everybody is dominant or subordinate in a relationship at some point in their lives, and we should accept changes in dominance and subordination as a fact of life, like growing up and getting old.

We must also accept the notion that the quality of life in a relationship, a family, a company, or a country may depend in large part on the personalities and the behavior of the dominant individuals within them. We can't prevent people from becoming dominant, but we can teach them that dominance comes with responsibilities. Dominants have leadership duties, and since subordinates pick up the tab for their success, dominants must make it easier on subordinates by being tolerant, generous, and forgiving. After all, dominance is not forever—we must be ready to step aside when the time comes.

Chapter 3

We Are All Mafiosi

Nepotism is something we can hardly do without. For one thing, nepotistic concern for the welfare of children is the engine of the capitalist system; take that away and you destroy the main incentives for innovation and the creation of wealth. For another, meritocracy unleavened by personal ties is inhumane, as ample evidence will show. Finally, on the individual level, nepotism is a profoundly moral relationship, one that transmits social and cultural values and forms a healthy bond between the generations. In short, nepotism works, it feels good, and it is generally the right thing to do. It has its origins in nature, has played a vital role in human social life, and boasts a record of impressive contributions to the progress of civilization.

—Adam Bellow, *In Praise of Nepotism: A Natural History*

The Raccomandazione

The Air Force base Caserma G. Romagnoli sits in Piazzale Aldo Moro, one of the busiest squares in Rome. The base is only a few blocks away from the Stazione Termini—Rome's main train station—and across the street from the campus of L'Università di Roma La Sapienza, one of the largest universities in Europe, home to over 100,000 students. The Caserma Romagnoli is an unusual base: no airplanes or helicopters can be found there, nor any military vehicles except for some blue passenger cars with the Air Force license plate. Instead, there are office buildings and a dormitory surrounded by tall perimeter walls. No one can see anything from the outside, and many college students who crowd the sidewalks along the perimeter walls don't even know what's inside.

It is seven in the morning on a cold winter day, and two young soldiers wearing the dark blue pants and light blue shirt of the Italian Air Force approach the main entry point to the Caserma, where a soldier standing in a booth monitors the pedestrian traffic in and out of the base. They appear to be carrying two heavy grocery bags. The guard in the booth asks to see the contents of the bags. One of them is full of raw meat: lamb racks, pork chops, ribs, steaks, and sausages. The other contains prescription drugs: antibiotics, pain relievers, anti-inflammatories, antidepressants, blood pressure and cholesterol drugs, and a multitude of others. The guard nods, snickers, and lets the two soldiers pass with their precious cargo. Clearly, he's seen this before and knows what it's about. I observe the whole scene from a few feet away and have no idea what's going on. Are the soldiers and officers on the base having a big barbecue? Did the infirmary run out of drugs? A few days later, after talking with other soldiers, I figure it all out.

To explain what happened at the entry point that day, first it's necessary to clarify the meaning of an important Italian word. The Italian word for "recommendation" is *raccomandazione*. "Recommendation" and "raccomandazione" sound similar, and according to the dictionary, they mean the same thing: advice, or support for an idea or cause. The two words are also used in similar contexts. For example, in both the United States and Italy, people applying for jobs may be recommended—or *raccomandati*—by someone else. This, however, is where the similarities end. In the United States, letters of recommendation provide an evaluation of a candidate's qualifications and are usually written by a senior person who is familiar with the candidate, such as a former teacher or employer. These letters are often a requirement of the application process; all candidates must have them; and although in theory they can be good or bad, in practice they tend to be uniformly good. As a result, good letters of recommendation don't necessarily increase one's chances of getting a job. Letters of recommendation make the most difference when they are bad.

In Italy, the raccomandazione is not a requirement of the job application process. It endorses a candidate but doesn't necessarily provide a description of his or her qualifications, it's usually made with a phone call instead of in a letter, and it generally comes from a family member or family friend. Not all candidates who apply for a job have a raccomandazione; those who don't generally don't stand a chance. For those who do have one, the chance of success depends not on how good the raccomandazione is—a raccomandazione can't be good or bad—but on the power and influence of the person who makes the call. The raccomandazioni are not meant to facilitate the review of applications by providing additional information about the candidates, but rather to rig the review process and guarantee the success of a particular candidate, regardless of his or her credentials. The raccomandazione is not advice or support; it's a request or even an order: make sure Mr. X gets the job. Typically Mr. X is a family member or a protégé of the recommender. The raccomandazione is the quintessential instrument for nepotistic influence on Italian public life.

Before we return to the Caserma Romagnoli and the mysterious dealings with meat and drugs that I witnessed, we also need some historical background. Until 1995, military service was mandatory in Italy. When Italian boys turned eighteen, they were drafted into a twelve-month service period in one of the armed forces. One day a young man would simply receive a postcard in the mail with information about his assignment. College students could request a deferral of military service to complete their studies. The military service could even be avoided altogether if one had a serious medical condition or was a conscientious objector.

In the 1980s, the decade when I turned eighteen, mandatory military service was not appealing to anyone. Italy was not at war, and there was no patriotic urge to enroll in the armed forces and protect the country. The military service provided no opportunity to earn money (the soldier's per diem was ridiculously low), to learn useful new skills, or to visit interesting locations. It was a major disruption of one's life that would be at best a gigantic waste of time and at worst a nightmare of stress and abuse. Being drafted into the Army and sent to a base in the mountains of northern Italy, for example, near the border between Italy and the former Yugoslavia, was considered the kiss of death. In the middle of nowhere, you had to march all day and stand guard all night, five nights a week. In addition, as a new recruit, you were physically, emotionally, and sexually harassed by the perversely named *nonni* (Italian for "grandparents")—senior soldiers near the end of their term—the way you might be harassed by career criminals in a prison in the Bronx. By comparison, service in the Air Force wasn't so bad. The Italian Air Force is small, poorly equipped, and has no real military power. Its main function is to provide logistical support to the Army and some civilian organizations. Many Air Force bases are located in nice urban areas, and soldiers do little or no military training. They can work office hours on the base and go home to sleep, are not required to march or be night guards, and are not harassed by nonni.

Draft assignments are supposedly made randomly. My friends and I used to joke that there must be a secret room in the Ministry of Defense in which a giant computer spits out draft postcards with random pairings between people's names and assignments, as in a lottery. Everyone knew, however, that the lottery was rigged. There were raccomandazioni. We imagined that every day a mysterious red telephone in the computer room rang and a voice on the other side said, "I recommend that Mario Rossi be permanently exempted from his military service due to a pathological heart condition" (as a result of raccomandazioni or a lack thereof, those with fake illnesses would be exempted, whereas those with real life-threatening conditions would be drafted), or, "I recommend that Dario Maestripieri be assigned to the Air Force and dispatched to a base in Rome." The gnome who answered the phone would then remove the name of the recommended individual from the lottery and make sure he got the recommended fate. Raccomandazioni would come in for thousands of young men every year. The other thousands of young men for whom the phone did not ring would simply get the default treatment: they were drafted and dispatched to the border. The family ties of these young men were torn, their education or careers were disrupted, and they were at risk of permanently impaired mental health from the effects of physical and psychological trauma.

The phone calls transmitting the raccomandazioni were probably made from offices on the upper floors of the Ministry of Defense—from high-ranking military officers who routinely gave orders to the gnomes in the computer room. They called to make sure their sons and nephews would either be exempt from service or else get the best possible treatment. These officers also called to recommend the sons and nephews of politicians, businesspeople, family friends, neighbors, or anybody who had a connection with them—or the power to make one.

After deferring my military service for four years to attend college, I began to panic. My parents were high school math teachers and had no direct connections with the military, the political parties, or big business. One day, however, one of my father's students—who was probably hoping to get a better grade in math—mentioned to him that her father knew an Army general who owed him a favor. My health was excellent, so trying to obtain a full exemption from the service on medical grounds had seemed a little too daring. I therefore told my father that I wished to do my service in the Air Force and be dispatched to Rome, where I would be stationed on the base across the street from the university and would have time to finish my thesis and spend nights at home. My wish was soon granted. The student's father called the general, the general made a raccomandazione on my behalf, and exactly one month later I received a postcard informing me that I had been drafted into the Air Force and would be stationed at the Caserma Romagnoli.

On my first day of service, however, I realized that something was wrong. I had expected that, given my college education, I would get a comfortable assignment in an office, but I found myself among the Air Force car drivers. This was very bad news. While the soldiers assigned to offices worked from nine to five and went home, the drivers worked late into the night taking generals back to their homes, sometimes hundreds of miles out of town. Something had gone wrong. My recommender had requested that I be drafted into the Air Force and dispatched to the base in Rome, but didn't specify a work assignment. As usual, in the absence of a raccomandazione, one receives the default, which means the worst possible treatment. At the Caserma Romagnoli, the worst possible job was being a driver.

There was more bad news. On that first day, the new drivers—a group of one hundred losers like me with "incomplete" or sloppy raccomandazioni—were unceremoniously informed that in one month ten of us would be transferred to an Army base in a bad part of town, where we would lose all of our Air Force privileges, including the reprieve from night guard duty and the opportunity to spend the nights at home. We were told that our training sergeant would randomly select ten drivers and forward their names to the captain in charge of the Caserma, who would then give orders for their immediate transfer to the Army base.

Everybody panicked. Within seconds of this announcement, we were all on the phone with our parents begging for a new raccomandazione that would save us from the transfer to hell. We initially thought that the raccomandazione would have to be made to the captain, but soon discovered that the sergeant wanted to be in on the game as well. The sergeant asked about our parents' professions and made a list of all of our names and our parents' jobs. He then hinted that if certain parents intervened on behalf of their sons, their sons would be spared from the transfer to the Army base. It became clear that he wasn't expecting to be bribed with money—he could have gotten into real trouble for that—but that he would appreciate other tangible favors. So much for random selection.

When I told the sergeant that my parents were high school teachers, my name quickly dropped to the bottom of his list. The soldier at the top had a father who owned a bank. The sergeant's son happened to be unemployed, so the soldier's father gave him a job in his bank. Another soldier had parents who owned a butcher shop. Going back to the earlier story, this was the young man with the steaks and sausages—he brought fresh meat to the sergeant almost every day for one month. The parents of another soldier—the companion of the butcher's son—owned a pharmacy and offered all kinds of expensive drugs to the sergeant free of charge. While it became clear that these three soldiers had nothing to worry about, things started looking very bleak for my lot. Thankfully I was eventually saved by another raccomandazione. A couple of phone calls were made, and the same general who got me into the Air Force "ordered" the captain at the Caserma Romagnoli not to transfer me to the Army base.

As these dramatic events were unfolding, I peeked through the window of the captain's office one day and glimpsed on his desk a list with the names of the one hundred new drivers who had recently arrived at the base, including mine. Next to each name, he had written in red ink the name of the military officer from whom he had received a raccomandazione. There was a lot of red ink on that list, and some drivers had received multiple raccomandazioni from different officers. A number of the recommenders were generals, and the captain had made a note of their rank. Clearly, answering the phone and taking notes on these raccomandazioni was an important aspect of the captain's job. Missing a call or pissing off the wrong general could have cost the captain his cushy position. As it was, he often left the base riding a powerful Ducati motorcycle and wearing black leather gear—not looking like he was going on any important military missions.

Luckily for me, next to my name on the captain's list was the name of a five-star general—I finally discovered my recommender's identity—so in the end I was saved from the unfortunate fate of being transferred to the Army. Not only that, but after a couple of more calls from my sponsor, I obtained a permanent release from all driving duties and spent the rest of my twelve months typing letters in an office. Every day at 5:00 P.M. I walked out of the Caserma and went to work on my thesis or to hang out with my friends. I slept at my parents' apartment every night and returned to the Caserma the next morning. Not everyone was so lucky. Sadly, there were ten names on the captain's list with little or no red ink next to them. These brave young men received the default treatment: they were transferred to the Army base and spent the remaining eleven months of their military service in hell.

Concorsi and Baroni

As illustrated by this story, the raccomandazione is a key instrument of Italian nepotism. To further demonstrate the inner workings of nepotism in my country I will use a different example, and once again, we start by learning some Italian words: *concorsi, baroni*, and *fregare*. *Concorsi* are nationwide competitions used both to admit college students into graduate programs and to hire new researchers and professors at public universities. *Baroni* (Italian for "barons") is a term used to refer to university professors who have great power and influence over student admissions, the hiring of new faculty, and funds for research. To *fregare* someone means to screw them.

Until 1980, Italian universities offered only one type of degree, called the *laurea*, which was a combination of a baccalaureate and a master's degree; then doctorate programs were introduced. To be considered for admission into a doctoral program, students had to compete in a concorso: their college grades and previous research accomplishments were evaluated, and they took an oral and a written examination. All positions were supported by fellowships, so students competed for both admission and money. This being Italy, the competition—you guessed it—was rigged. The baroni negotiated with one another the number of students they could each admit each year into their programs and who would win the concorsi. Before applications for admission were even received, the baroni would have already decided on the winners. Others were told not to apply and either wait their turn or simply give up the whole thing.

When it came to admission decisions, family, of course, came first. The baroni admitted their children and other family members directly into their programs or recommended them for admission to other baroni. Baroni also guaranteed admission to their protégés. These were undergraduate students who, thanks to a raccomandazione from their parents, had completed their undergraduate thesis with a barone and, because of their loyalty to him, had been granted the status of extended kin—they had been adopted. Finally, the baroni admitted students who were neither their kin nor their protégés, but strangers for whom they had received raccomandazioni from politicians, businesspeople, or friends and neighbors. As usual, all raccomandazioni were made over the phone so that no traces would remain. Any student who applied for admission but did not fall into the above categories would be turned down by the baroni regardless of his or her academic credentials. My adviser turned down a lot of students without the required family pedigree or raccomandazioni even if they were academically outstanding and he had empty slots in his lab. He had to keep them vacant because his phone could ring at any time with a request to take a student he couldn't turn down. As a result, most of the students and researchers who worked with him were the sons and daughters of other professors or politicians. So how did I get in there?

The year I applied for admission into the biology doctorate program at the University of Rome there were eight open slots, and as usual the names of the eight winners had already been agreed upon by the baroni on the admission committee. Mine wasn't one of them. A couple of weeks before the concorso, however, the National Research Council offered funding to support two additional fellowships. The baroni did not have time to negotiate these two additional positions, so two outsiders who had good résumés and had performed well on the exams—myself and a friend of mine—were admitted. We squeezed in through a crack in the system. Then something funny happened. Even though my friend and I were straight-A students and had already published scientific articles, we couldn't find a professor who was willing to serve as our adviser. The day the new graduate students were to be assigned to an adviser we sat around a table with a bunch of professors, and each and every one of them, with a great display of embarrassment, made excuses as to why he couldn't serve as our adviser.

The truth was that by filling a slot with an outsider without raccomandazioni or the appropriate family pedigree, they might lose an opportunity to admit a family member or the son or daughter of the prime minister into their program the following year. It now became apparent to the baroni that admitting two outsiders into the program had been a big mistake—someone would have to pay the price. Eventually, after some arm-twisting, my friend and I found an adviser. Three years later, however, after I finished my PhD, it was made abundantly clear to me that someone who had entered academia through a crack in the system could not expect to go very far. After doors were shut in my face one too many times, I packed my bags and moved to the United States.

The PhD students who enter the system normally, through the proper channels, remain in their advisers' shadow for years until their loyalty is finally rewarded with a permanent position. During these years the protégés spend as little time away from their patrons as possible because their future careers depend on the strength of their personal bonds with them, and they know these bonds must be constantly attended to and nurtured.

The nepotism that controls admission to graduate programs is nothing compared to what happens when academic jobs and real money are involved. It is common knowledge that all concorsi for full-time researchers and professors at Italian universities, and especially in medical schools, are rigged by the baroni; complaints and appeals by candidates who were unfairly turned down for positions for which they were eminently qualified (in a word, they were *fregati*) have led to countless criminal investigations and even some convictions for the baroni involved. From these investigations of academic nepotism, it has emerged that the baroni have organized themselves in clans that operate just like the Mafia. They have hierarchies of power with a "boss" at the top, they aim to control entire areas of academia across the country, and they do not hesitate to threaten and intimidate to get what they want. Scandals involving rigged concorsi have received a great deal of media attention in Italy; countless newspaper and magazine articles, and even books, have been written on the subject. Several years ago, the weekly news magazine *L'Espresso* devoted a cover article—entitled "The Baroni's Mafia"—to academic nepotism in Italy and reviewed some of the best-known scandals.[1]

Among the incidents recounted in the article, a few are particularly noteworthy. For example, twenty-five new professor positions in otorinolangoiatry were filled in universities around Italy in 1988 and 1992. Of these twenty-five new hires, four were the sons of professors who sat on the search committees that examined the candidates. One powerful barone, Dr. Giovanni Motta, appointed his own son, Dr. Gaetano Motta, as a full professor at the age of thirty-two. The father—as the chair of the search committee—himself evaluated his son's credentials, which included scientific articles published in his father's department, with his father as a co-author. The senior Dr. Motta then falsified the reports of the examinations to make it look like his son was more qualified and had performed better on the examinations than the other candidates. Dr. Motta and other baroni whose sons were hired in these concorsi were later found guilty of fraud and sentenced to one to two years in jail. Although the hirings were declared null, Dr. Gaetano Motta to this day still holds the professor appointment he illegally obtained in 1992.

Another case that made the news involved Dr. Roberto Puxeddu, an associate professor at the University of Cagliari. He was appointed by a committee that included two professors who had themselves obtained their faculty positions through a fraudulent concorso chaired by Prof. Paolo Puxeddu, Dr. Roberto Puxeddu's father and a powerful barone. Again, although the senior barone was later convicted of fraud and his son's appointment annulled, the son maintains his position at the university. In another case in the medical school of the University of Bari, a professor who became dean left the directorship of his department to his thirty-four-year-old son, who was the only candidate considered for the position. Another dean pressured his university to hire his daughter without even advertising the position and interviewing other candidates.

The inner workings of the Italian academic mafia were revealed when some university phones were wiretapped and conversations between baroni were recorded by the police. In 2005, Dr. Paolo Rizzon, a professor at the University of Bari, was recorded discussing strategies for manipulating concorsi across Italy. In one such conversation, he negotiated the composition of a favorable search committee for his son, who had applied for a faculty position, and then he negotiated the topic of the essay his son would have to write as part of his examination. Another recorded conversation revealed that a qualified job candidate who competed against the baroni's protégés was threatened with physical violence by two Mafia hit men if he didn't withdraw from the concorso. The two hit men were identified by name—both had criminal records. In another recorded conversation, Dr. Rizzon bragged to a colleague that in order to help his son and the relatives of other baroni obtain professorships, he had to be very creative to be able to fregare outsider job candidates whose qualifications were much better than those of the people he helped.

The qualified job candidates who are fregati by the baroni often leave the country and begin successful careers abroad. In the last twenty to thirty years, tens of thousands of Italian researchers have fled the country. The baroni's clans continue to operate undisturbed and have absolute control of the Italian academic system. As a result of such nepotism, the Department of Economics at the University of Bari had, at some point, eight professors who shared the same last name: Massari. They were all related. Apparently, this set a new record for Italy; the previous record was six family members in the same department or institution. Of course, when it comes to nepotism, Italian military officers or the baroni of academia are amateurs when compared to politicians, judges, businesspeople, and anyone else who has real power and influence in society.

In his book *In Praise of Nepotism*, Adam Bellow—himself the son of American novelist Saul Bellow, who won the Nobel Prize for Literature in 1976—describes outrageous cases of nepotism that have come to the media's attention around the world.[2] You might ask, what has Bellow found to praise about nepotism? Before I address this question, I first explore his assertion that the origins of nepotism lie in nature.

The Nature of Animal and Human Nepotism

To a biologist, nepotism simply means favoritism toward kin, such that kin are preferred as social (but not sexual) partners and helped at the expense of nonkin. For example, a nepotistic squirrel who has saved a couple of nuts for dinner will share one of them with his starving brother but not with the unrelated squirrel next door. Put another way, biologically speaking, nepotism is altruism directed toward family members. This altruism, however, is a bit phony. Since family members share some of their genes, a nepotist is maintaining his or her own DNA in the population by helping a relative. So nepotism is really selfishness in disguise. Many selfish behaviors have evolved by natural selection because they help an individual to survive and reproduce; the genes for selfishness are transmitted to the next generation through the selfish individual's own children. Similarly, many nepotistic behaviors have evolved through a particular kind of natural selection, called kin selection, because these behaviors help an individual's relatives to survive and reproduce; the genes for nepotism are transmitted to the next generation through the nepotistic individual's family members.[3]

Nepotism is a universal phenomenon. Animal species vary in levels of nepotism, just as human societies do. But there is no animal species or human society in which individuals are biased in favor of nonkin against their kin. What makes animals or humans more or less nepotistic is usually the availability of resources. When everyone has all the food (or water or money) they need or want, they can afford to be generous with others and don't bother to discriminate as much between kin and nonkin. When belt-tightening becomes necessary, however, family values rise in importance. Now, it's not very often that people have all the money they want—which may explain why nepotism has been part of human history since Adam and Eve were kicked out of the Garden of Eden. Since then, humans have had to work hard to earn their bread and butter, and in so doing, they have always helped their family members against nonrelatives. There are so many examples of nepotism in the Bible that you might call it the bible of nepotism.

But animals had been behaving nepotistically long before Adam and Eve. For example, social insects such as ants, bees, and wasps have been practicing this behavior for millions and millions of years.

Their societies rely heavily on cooperation within groups of relatives and aggressive competition against groups of nonrelatives. Examples of nepotistic behavior can be found in almost any animal species, from vampire bats, who regurgitate the blood of their victims only to their close relatives, to naked mole rats, burrowing rodents native to East Africa, among which many females give up sex altogether to perform hard labor such as digging tunnels and gathering food for their mother, the queen. Some species of monkeys and apes that are closely related to us have taken nepotism to the next level. They don't simply help their relatives with food but also help them gain and maintain political power. One of the most political and shamelessly nepotistic creatures on this planet is the rhesus macaque, a monkey species I happen to be extremely familiar with.[4]

Like humans, rhesus macaques live in a very competitive society, and not surprisingly, rhesus macaques, too, are obsessed with dominance. As discussed in Chapter 2, dominance between two rhesus macaques is established on the basis of asymmetries in their resource holding potential. The rhesus RHP, however, looks more like that of U.S. congressmen, for example, than that of an animal species such as deer, among which a male with larger antlers becomes dominant over a male with small ones. A Washington politician's RHP has nothing to do with how large his antlers are, but everything to do with how much political support he has from his party and how powerful the party is. The same goes for rhesus macaques. When it comes to dominance, the only asymmetries that matter are those of political support. Adult females and juveniles receive support mainly from their family members, and this support takes the form of agonistic aid. When the daughter of the alpha female picks a fight with another adolescent female from another family, the alpha female and her sisters immediately join the fight and help their young relative defeat the other adolescent and her relatives. Similarly, what happens when Tony Soprano's nephew wants to gain control of the drug-dealing in his neighborhood? His uncle sends a couple of hit men to whack the competition. Clearly, human nepotism has its origins in nature. In fact, rhesus nepotism and human nepotism seem exactly the same. But are they really?

Let's take a closer look at the nepotistic monkey business. Rhesus macaques live in a matriarchal society. In a rhesus group, there are several families, but these families do not consist of a mother, a father, and their children. The main contribution rhesus males make to their descendants is the sperm with which they impregnate the mothers of their children. They try to be as generous as possible with their sperm donations; when a dozen infants are born six months later, fathers simply disappear from the scene to go hang out with their buddies at the casino. Because of this father absenteeism, rhesus families consist of multigenerational groups of female relatives, called matrilines, with their young offspring. For example, a typical matriline may include a ten-year-old adult female with her mother, grandmother, sisters, aunts, cousins, offspring and grand-offspring, and nieces and nephews. The males are part of the family until they reach puberty at around five years of age; then they wave good-bye to everyone and emigrate to another group. The females stay attached to their mother's apron strings forever.

The matrilines within a rhesus group have power in the same way that political parties or the Corleone and Soprano families do. The more numerous the members of a matriline, the greater the power. To give an example, in a rhesus group there could be three matrilines ranked in a hierarchy of power. The largest matriline is at the top, the smallest at the bottom, and the third one in the middle. The power of the matrilines is transferred from the older to the younger monkeys through nepotistic intervention in agonistic confrontations. Since juveniles pick fights all the time with other juveniles, as well as with adults, and their mothers continue to intervene on their behalf, the sons and daughters of high-ranking mothers become more and more powerful and eventually acquire a dominance rank just below that of their mothers. The sons and daughters of low-ranking mothers also end up with a rank similar to that of their mothers, which means that, unfortunately for them, they become losers like their family members.

In the rhesus society, destiny is set at birth, at least for females. The females that are born into monkey aristocracy grow up to be more and more aristocratic thanks to the nepotism of their mothers and other relatives. Those that are born with low status prepare themselves for a life of misery. But there are females who are even worse off than those belonging to the lowest-ranking matriline. If a female is abandoned at birth by her mother, rescued and raised by a compassionate researcher, and then reintroduced into the group a few years later, this female will be a Cinderella without a family. Cinderellas don't fare well in nepotistic systems. If a rhesus Cinderella somehow manages to survive in this hostile environment, she might get lucky and be impregnated by a male of low rank. Certainly no Prince Charming will ever take her to the ball and turn her into a princess. But she will have a chance to start a family of her own, and a few generations down the line, if she and her relatives can survive the burden of doing chores all the time without having any fun, the family may become large enough to be able to claim a better place in society.

We've seen that in rhesus macaques nepotism transmits social status and strong bonds between family members across generations the way it happens in many human societies. But again, are monkey nepotism and human nepotism exactly the same thing?

Animal nepotism and human nepotism differ in several important respects. For the sake of consistency (and because I really like these monkeys), let's continue using rhesus macaques as an example. Nepotism in macaques is mostly a female business, and especially a maternal business. Males don't recognize their offspring, don't give them milk bottles or change their diapers (or their monkey equivalent), and don't help them realize their dreams of wealth and world domination the way human fathers try to do with their children. So the first important difference between humans and rhesus macaques (and other animal species as well) is that in humans nepotism is very much a male business. Traditionally, in human societies, men have always held most of the wealth and political power, so it's not surprising that men's behavior can be highly nepotistic. Matriarchal societies are not infrequent in mammals; for example, elephants behave very much like rhesus macaques. Humans are unique among animals for the prominence of the patriarchal family and the strength of paternal nepotism. In my personal experience in the Italian military and in academia, it was always fathers, and rarely mothers, who pulled strings on behalf of their children.

Another important difference between rhesus and human nepotism is that while in the monkeys nepotism is limited to biological relatives, in humans it isn't. Humans have extended the boundaries of the biological family to include nonkin in two ways: through marriage and through patronage.[5] When we marry, we agree to treat our spouse and our spouse's relatives as if they were genetic relatives. Throughout human history, marriages and exchanges of wives have also allowed men to form nepotistic alliances with men from other villages or tribes. In humans, just as in rhesus macaques, political strength lies in numbers. For men and women with strong political ambitions, even an extended family may not be enough. Nonrelatives, then, must be brought into the family and given kin status. The Mafia provides a good example of this phenomenon. The Mafiosi maintain strong bonds with their relatives but increase the size and power of their families by providing patronage to a large number of associates. The head of the family cements this patronage by serving as a godfather to the children of these associates.

All human societies have developed means of incorporating strangers by according them the status of relatives. In the Italian academic system, the barone offers patronage to the students and researchers who have made it into his feudal castle through appropriate raccomandazioni. In exchange, he expects subordination, loyalty, and in some cases even bribes or sexual favors. A few years ago, Ezio Capizzano, a sixty-six-year-old law professor at the University of Camerino, made the national headlines in Italy because it turned out that for years he had been having sex with his female students on a couch in his office and making videotapes with a camera hidden under the table.[6] The students in his tapes always passed his course exam with flying colors. The professor even negotiated good grades for his young lovers with other baroni, when the students had to take courses with them. When he was caught, Professor Capizzano claimed in his defense that his sexual relations with the students were all fully consensual. Judging from the photos in the newspapers, I didn't get the sense that he was particularly attractive, and it's likely that his female students had sex with him for business purposes: they exchanged sex for good grades and raccomandazioni.

The Mafia families are really no different from the dynastic families that ruled various countries around the world for centuries. In his book, Bellow nicely illustrates that much of human history is the history of the rise and fall of dynastic families. He also notes that humans even extend kinship status to the gods. All primitive religions involve the formation of kinship relations with the gods in hopes that they will treat their followers favorably, and the leaders of these religions typically consider themselves to be direct descendants of these same gods. By offering food and making sacrifices, people attempt to establish a patronage relationship with the gods to make them feel obliged to reciprocate these gifts with favors or support.

Another difference between rhesus and human nepotism is that while the monkeys transfer only their social status to their relatives—some other animals transfer nests or territories to their offspring—humans transfer not only their power and privileges but also their property, money, knowledge, and values to their descendants. Therefore, human nepotism is also a cultural phenomenon, since the intergenerational transmission of knowledge, norms, and values within families makes an important contribution to human cultures. The trouble with human nepotism, however, is not that relatives are educated or that they are helped, but *how* they are helped. This is the source of the most important difference between the nepotism of rhesus macaques and our own. It has to do with a thing called morality.

Like everything in nature, rhesus macaque nepotism—and all animal nepotism—is neither good nor bad. Sure, there are winners and losers: in the rhesus world, high-ranking females are winners and low-ranking ones are losers; in the African savanna, the lion that captures the gazelle is the winner, and the gazelle that ends up in its stomach the loser. But the lion is not a bad animal, nor is eating the gazelle wrong.

Humans started out their evolutionary journey similarly amoral. Then Moses gave people the Ten Commandments and told them what was good and bad, and right and wrong, according to God. Or maybe a bunch of people just sat around a table and signed a social contract establishing some rules for peaceful coexistence in human societies. The day the contract was signed, human nepotism diverged from animal nepotism. High-ranking rhesus macaques torment and torture unrelated monkeys of lower rank, but in doing so they don't break any rules. Moses never spoke to the monkeys about commandments, and the monkeys can't write a social contract—or they haven't had time yet to do so.

Allow me a brief digression. If the "infinite monkey" theorem were correct, given enough time rhesus macaques would produce some sort of social contract with norms and rules for their society. The theorem states that monkeys hitting keys at random on a typewriter keyboard for an infinite amount of time should be able to produce the entire works of William Shakespeare. However, this theorem, or at least a simplified version of it, was disproved by an ingenious experiment conducted at the Paignton Zoo in England. Zookeepers left a computer keyboard in the cage of six macaques for a month.[7] Contrary to the theorem, the monkeys produced only a five-page document, consisting mostly of the letter S, until the alpha male bashed the keyboard with a stone and all the other monkeys urinated and defecated on it. So no contract and no morality for the monkeys.

When people behave nepotistically in public life, they almost invariably break moral, social, and legal rules. If everybody played by the rules, nepotism would be useless. Moral inclinations are strong—in some individuals more than in others—but the instinct to favor relatives is even stronger. In the end, rules are broken all the time, and nepotism has ended up being associated with fraud, corruption, and a host of other crimes. The popes in Rome, instead of playing by the rules and appointing people to office based on merit and qualifications, hired their illegitimate sons, whom they described as "nephews"—hence the term *nepotism*. In doing so, they had to *fregare* more qualified individuals, just as Prof. Paolo Rizzon did to give his son an undeserved professorship. If being *fregato* was the only consequence of being a victim of nepotism, things wouldn't be so bad (although being sent to an army base at the border between Italy and the former Yugoslavia or not getting a job can destroy a person's life). The problem is that fraud is the least of the crimes that have been associated with nepotism throughout human history. Millions of people have been killed as a result of the nepotistic behavior of ruthless dictators bent on advancing the interests of their family members at all costs. Uday and Qusay Hussein—the two sons of Saddam Hussein, who were killed in a gun battle with U.S. forces in 2003—would not have acquired their immense power and wealth without their father's ruthless support and the shedding of hundreds of Iraqi citizens' blood. Criminal nepotism is still rampant in many human societies, and particularly in dictatorships in Africa, Asia, and South America. According to Bellow, Europeans, too, have a relatively positive and tolerant view of nepotism. American society, by contrast, was founded on the criteria of merit, fairness, and equal opportunity, and Americans have historically shown resistance to and rejection of nepotism.

Just as America has seemed to succeed in producing the best meritocratic society in the world, however, nepotism has reemerged stronger than ever. Throughout the twentieth century and up to the present day, we have seen the success of family dynasties in politics, business, arts, music, and literature. These families have employed the nepotistic strategies used by European elites to transfer their wealth and power to successive generations. Capitalism is supposed to be the great engine of individual liberation from the confines of the family. Yet family interests continue to predominate in American economic life. In fact, in a new book titled *A Capitalism for the People: Recapturing the Lost Genius of American Prosperity*, Luigi Zingales— an economist at the University of Chicago Business School and a colleague of mine—makes the case that American capitalism, once unique in the world for being based on fair competition, equal opportunity, and meritocracy, has been gradually changing and increasingly resembles Italian capitalism, in which cronyism and nepotism rule.[8]

Powerful families protect their interests in their personal lives as well. Members of the upper class—whether they're the political, business, or intellectual elite—increasingly live in their own segregated communities, send their children to the same exclusive schools, marry individuals of their own class, and act in other ways to pass on their wealth, power, and privileges. I can personally attest that nepotism has been creeping into American academia, increasingly following patterns that are similar to those of Italian baronism. I moved to the United States in 1992, and of the first two academic jobs I interviewed for, one was offered to the daughter of a powerful professor in the same institution, while the other was offered to an internal candidate, a protégé of the department chair. At my own institution, the University of Chicago, many of the students supervised by well-established professors happen to also be the sons and daughters of other well-established professors. In fact, as I was writing this chapter, I received the following email from a colleague:

> Dario,
> We met briefly last week. I am Vice-Chair of the Department and I was very impressed with the work you are doing.
> The reason I am writing you is to ask if you have an interest in an intern for this summer. My son is——at the University of——and we have been talking about your work. Would you be willing to look at his CV and a possible internship for the summer? I would appreciate anything you can do to help him get this experience.

Sounds like an Italian raccomandazione, doesn't it? Maybe this is my chance to get a lamb rack or some Italian sausages.

As I mentioned earlier, biologists explain the relative strength or weakness of nepotism in a species or a society on the basis of availability of resources and the intensity of competition. With America's recent history of resource depletion and economic crises, coupled with healthy population growth, social competition has intensified. America may still be the land of equal opportunities, but some are more equal than others. At the present time, a great deal of wealth and political power is concentrated in the hands of baby boomers: people born between 1945 and 1960. And as the baby boomers approach retirement age, their children are entering the workforce en masse. Sounds like the perfect setup for nepotism, doesn't it? No wonder the aging baby boomers are using all the means at their disposal to transfer their wealth and power to their children.

Confronted with the reality that American society is becoming less meritocratic and more nepotistic, Bellow launches himself into a patriotic defense of American nepotism. He explains to us that contemporary American nepotism is a different beast, of a gentle and noble kind, and nothing like the nepotism practiced by rhesus macaques, dictators around the world, or Europeans. Therefore, he reassures us, any concerns we might have about the resurgence of nepotism in America are unfounded. He also tells us that while bad nepotism is essentially harmless, good nepotism plays a positive role in promoting a capitalistic economy and conservative moral and family values.

The Good, the Bad, and the Ugly

Although nepotism is traditionally viewed as a top-down phenomenon—for example, parents pulling strings to help their children—the new American nepotism also has a significant bottom-up component. It is often the result of children choosing to follow in their parents' footsteps and thus inheriting jobs, wealth, and political power, rather than of parents bending rules to get jobs for them. The opportunism of children who exploit the family name and their parents' wealth and influence is not limited to the upper class but is practiced by anyone who has connections. There is nothing new, however, about this bottom-up aspect of nepotism. In fact, the top-down and bottom-up components of nepotism have always gone hand in hand. When children become aware that they live in a nepotistic world, learn what nepotism can do for them, and realize that their parents have money, power, and influence, they actively try to take advantage of every opportunity for gain that is available to them. The beneficiaries of nepotism are never passive bystanders. The same is true in monkey societies. Young rhesus macaques born to high-ranking mothers actively solicit their mothers' interventions by continuously picking fights with others and screaming for help.

Bellow acknowledges that the resurgence of interest among baby boomers in creating dynastic genealogies by transferring business, wealth, and power to their children has also been accompanied by a reversion to older expressions of nepotism. He gives this example. Richard Williams one day announced to his wife that he was going to turn their daughters—named Venus and Serena—into tennis champions. So he taught the girls tennis as soon as they could hold a racket in their hand, and he continued to give them lessons until they won their first U.S. Open. Says Bellow: "As a father, Williams represents what can only be called a throwback to Ancient Nepotism. His example is a testament to the nepotistic truth that where children are concerned, you get out what you put in." Well, if this is a good example of a resurgent Ancient Nepotism, we've got nothing to worry about.

Teaching your children how to play tennis is not nepotism. And more generally, there is nothing wrong with parents who transfer their skills, business, or money to their children. Nepotism occurs when individuals, in the process of helping their relatives or protégés, break the rules. And the more parents try to influence their children's lives, the greater the chance they will break the rules. Aging baby boomers are becoming involved in the lives of their children to a much greater extent than their parents ever did with them. And yes, this phenomenon has been accompanied by a resurgence of nepotism, but there is nothing gentle or kind—or new for that matter—about it. It's the same old nasty beast. To be fair, nepotism is increasingly practiced not only by members of the upper class but by members of the lower class as well. Nepotism, however, is much more dangerous to society when rich and powerful people do it. The lower classes simply don't have the power to bend the rules, and their nepotistic activities are largely inconsequential for the society.

It is possible to define nepotism very narrowly to mean not just hiring a relative, but hiring one who is grossly incompetent over more qualified candidates. Here, Bellow introduces the crucial distinction between bad and good nepotism. Bad nepotism, such as hiring an incompetent relative, typically backfires, but it's a relatively minor offense because it doesn't hurt people other than those directly involved in it. By contrast, good nepotism, which involves helping a competent relative, arguably benefits everyone.

So Bellow's conclusion is that "history shows that nepotism in itself is neither good nor bad: how it is practiced is what matters." We shouldn't eliminate nepotism or punish it, but rather apply it constructively and recognize that it's an art. Moreover, we must observe the unwritten rules that have made it a constructive and positive force in our society. Those who observe these hidden rules are rewarded and praised; those who don't are punished. And—drum roll, please—these are the golden rules:

First, *don't embarrass the patron*. Since the protégé's actions and conduct reflect on the patron, accept the nepotistic help but don't make the patron look bad. Second, *don't embarrass yourself*. If you are the beneficiary of nepotistic help at someone else's expense, you must work hard to counteract the resentment of those who are fregati. If possible, you should give them a consolation prize. Third, *pass it on*. If you are the recipient of nepotistic help from your parents, express your gratitude to them in the form of nepotism toward your own children. It is okay to receive generous nepotistic help if you become, in turn, generous toward others (as long as you keep it within the family, of course).

Mafia Family Values

Despite providing reassurance that this is not the case, Bellow reduces the subject of nepotism to an argument about family values. He tells us that we have a moral obligation to be nepotistic: if we fail to put our families first, we may destroy the very fabric of human society. We should therefore strengthen nuclear families, encourage people to help their relatives, and stimulate the creation of extended kinship networks through patronage of friends and associates. Hiring a nephew may be objectively discriminatory, but since people are going to do it anyway, we may as well make sure that we hire the best and most meritorious of our nephews, concludes Bellow.

"If nepotism is just about helping relatives," Bellow reasons, "then clearly there is nothing wrong with it and even the nepotistic values the mafia embodies may have merit and legitimacy." He cites an episode of the TV series *The Sopranos* in which Tony Soprano's wife Carmela tries to get their daughter admitted to Brown University by pulling strings. She says: "It's all connections now. It's who you know. If the rules don't apply to everyone, why follow the rules?" So, if we share Carmela Soprano's and Adam Bellow's views, it can fairly be concluded that, in the end, we are all mafiosi.

Chapter 4

Climbing the Ladder

Going It Alone

Nepotistic raccomandazioni from our biological or adoptive parents—such as baroni, army generals, or politicians with whom we have connections—are one way to advance our careers and secure good and comfortable lifestyles. Sometimes, however, nepotistic support is not an option. For most of us, there are times in our lives when we must leave our families and join a new group in which we have no connections to help us. Chances are that this new group already has a clear power structure and a well-established dominance ladder. Our success in this new environment will depend on our ability to negotiate the ladder, and we have no one to depend on but ourselves.

Male rhesus macaques on the brink of adulthood are confronted with this very situation. When male macaques reach puberty, they emigrate from their familial group—in which their friends and their mother, sisters, and other maternal relatives reside—and join a new, entirely unfamiliar group. Like a young male rhesus macaque who realizes he can't make a reproductive career in his natal group, I left my family, my support network, and my country after college to emigrate to the United States. In both macaques and people, such transfers are by no means limited to the beginnings of reproductive or professional careers. Monkeys and people also make "secondary" transfers later in life in order to seek even better reproductive and career opportunities. The kinds of problems they encounter in these secondary transfers, however, are generally quite similar to those of their first move.

Whether you are a rhesus macaque or a human being, you've probably experienced transfers into a new group at some point in your life. And you know that not everyone will roll out the red carpet for a newcomer. Even though most people who enter a new workplace have been hired, which means that they are wanted and welcomed by someone in the organization, they must contend there with an established power structure that is generally resistant to change. As discussed in Chapter 2, human workplaces—whether they are large corporations, military organizations, theater companies, or schools and colleges—have dominance hierarchies, just like monkey groups. People who have worked hard to climb the ladder—whether they are now all the way at the top or simply one step up from the bottom— are not happy to step aside and make room for a newcomer. In both macaque and human societies, the newcomer is seen as a competitor, and therefore his or her arrival into the group is likely to be met with indifference, resistance, or outright rejection. The newcomer, then, must cultivate relationships with these strangers and try to obtain their support through exchanges of favors or other means. When there are no relatives around to help, success depends not on nepotism but on politics. But in humans and macaques, there are different ways to play the politics game, and different ways to climb the ladder.

Here I would like to explore three different strategies for climbing the ladder that are used by both humans and male macaques. Each strategy is illustrated by a story: I begin by sharing the three human stories—with fictional characters named Gina, Mario, and Sarah— and then tell three parallel monkey stories involving male macaques named Billy, Rambo, and Max.

Primatologists have developed theoretical models that explain why these different strategies for climbing the ladder exist and under what circumstances each one is most effective. Primatologists have also created models that explain when it is advantageous to form political coalitions with others to change the order on the ladder and what kinds of coalitions are most advantageous under what circumstances and why. As we've seen before, these models combine principles from evolutionary biology with economic cost-benefit analysis. Although the way people and monkeys play the game of politics is not exactly the same, there may be enough similarities between the people's stories and the monkeys' stories to convince you that the theories that explain monkey behavior can explain our behavior too.

Human Stories

THE GOOD CITIZEN

In her first few years of employment as a systems analyst at Microsoft, twenty-seven-year-old Gina worked hard and kept a low profile. Whenever she ran into a coworker in the hallways of their building, she would smile and make small talk. During business meetings in her department, Gina sat quietly in the back of the room and never asked questions or volunteered an opinion on the topics of discussion. When a coworker turned to her for a reaction, she would smile and nod. If directly asked for an endorsement of a proposal that the majority supported, she readily offered it. Asked for an opinion on a controversial subject over which the group was divided, she would diplomatically avoid taking a position on the issue, claiming that she was new to the company and didn't know enough about the matter in question to have an opinion. She never missed a meeting and regularly showed up at the department's "happy hour" events.

Gina was often called into her boss's office and given extra assignments. Thus, in addition to the normal duties that were part of her position, she also took on extra work for the department that was above and beyond her responsibilities. When the boss would ask in a business meeting if anyone was willing or had time to take on some extra work, everybody else made excuses about being too busy or not having the expertise or the resources to take up the new responsibility. Gina, however, made no excuses and readily volunteered for the new assignments. And it wasn't only the boss who was giving Gina extra work. Other senior colleagues of Gina's, by asking her for this or that favor, also dumped on her aspects of their work they didn't like. Gina, of course, couldn't say no. Gina wasn't the only employee who was asked to do extra work; this happened to other junior employees who had been with the company for a year or two. Like Gina, her junior coworkers couldn't dodge requests to do extra work, but they didn't look as pleased as she did when accepting the extra assignments, and in some cases they voiced a complaint. They sometimes missed business meetings or happy hours not only to avoid getting caught in these situations but also to register a silent protest against what they considered "professional abuse" on the part of their boss and their senior coworkers. This, of course, meant that Gina—who never missed a meeting—ended up doing most of the extra work.

Behind the facade of smiles and happiness, Gina was quite resentful about the additional work she was forced to do. Although many requests came from her senior coworkers, Gina wasn't as mad at them as she was at her junior colleagues who, by not showing up at meetings, avoided some of these requests. Although Gina and her junior colleagues were essentially in the same boat and could have benefited from establishing friendly and cooperative relationships among themselves, Gina didn't think that it would advance her career to form alliances with coworkers who appeared to be as powerless as she was. She decided that the best—or the only—opportunities for improving her professional position would come from the support and generosity of the people at the top of the power hierarchy. Gina counted on the expectation that by constantly pleasing her superiors and accommodating their requests she would eventually be rewarded by them. She knew that it was highly unlikely that her boss would offer her a big promotion or a significant salary raise out of the blue. But she thought there was a chance that being a good citizen and keeping a low profile would enable her to experience a slow but gradual rise through the ranks. It helped Gina in her use of this strategy that she had an easygoing personality and a nonreactive, unimpulsive temperament. She knew that making it in this new workplace would take a long time, but she was willing to wait—without rocking the boat.

The Young Turk

Mario was a twenty-six-year-old biologist who had just finished his doctorate and accepted a two-year position as a postdoctoral researcher at the University of Arizona in a research group led by Prof. Michael Levine, a world-renowned authority in his field. Mario had been a straight-A student from first grade all the way through graduate school and never had a doubt in his mind about his potential for a successful career in academia. During his years in graduate school, he had done a great deal of research, sharpened his critical thinking skills, and developed the ability—or so he thought—to distinguish interesting research questions from uninteresting ones, to tell good science from bad, and to discriminate between smart scientists and phonies. Despite the fact that Mario had only recently obtained his doctorate, he had already achieved some recognition as an independent scientist by publishing articles and being invited to give presentations at prestigious conferences.

Twenty years of successful performances, first in school and later in college, had given Mario a high degree of self-confidence. This belief in the value of his own ideas and the quality of his work had also produced an arrogance in Mario, accompanied by a disregard for others' viewpoints. Mario was a highly driven and motivated young man who pursued his goals with extreme tenacity and persistence. Not surprisingly, he was also a self-absorbed person with poor social skills—he was embarrassingly bad at reading other people's emotions and communicating with them without hurting their feelings. A highly competitive and anxious individual—traits commonly associated with high testosterone in the body and low serotonin in the brain—Mario was also a highly impulsive person who did not hesitate to launch himself into direct confrontations with others, at least on intellectual matters.

When Mario joined Dr. Levine's research group, he was familiar with his new boss's previous research but hadn't yet formed a full opinion of him professionally, nor did Mario know anything about Dr. Levine personally. In his first few weeks in Dr. Levine's group,

Mario concentrated on his work and on being productive and made no efforts to socialize with other members of the research group. Not surprisingly, he came across as antisocial, aloof, and socially challenged. Having always been a one-man show when it came to doing research, Mario made no effort to establish collaborations with the other researchers and students. Although he was familiar with and admired the work done by some of these people, his failure to communicate with them and let them know that he appreciated their work made them feel that he looked down on it. They saw Mario's failure to engage with others as an expression of arrogance. As a result, he was disliked by everyone in the group and had no support. For Mario, being socially challenged, this wasn't a problem: when he found himself in a group of "friends," half the time he couldn't tell whether anyone liked him or not, and the rest of the time he knew that no one liked him but didn't care. Moreover, when it came to school and later to work, he was firmly convinced that his potential for success depended entirely on himself, his skills, his dedication, and the quality of his work. He didn't think that networking or friendships had anything to do with advancing his career. He also had little respect for authority derived from prestige and political power and couldn't stand people who pulled rank on him. He believed that in the intellectual arena authority must be earned from confrontations of ideas and that anyone could—and should—be challenged.

And so virtually from the day he started working in Dr. Levine's group, Mario began challenging his boss in endless discussions of their research, citing the strengths and weaknesses of Dr. Levine's articles versus Mario's own papers. He put himself on the same intellectual level as Dr. Levine and tested the strength and quality of his own ideas and work directly against those of Dr. Levine. Needless to say, Dr. Levine didn't like this at all. He thought that although this young man was somewhat smart and knowledgeable, he was far too opinionated and self-assured given the tenuousness of his previous experience and accomplishments. Although Dr. Levine occasionally thought that the discussions with Mario were interesting, most of the time he was annoyed. He had no interest in being intellectually stimulated if he wasn't going to get the respect a senior researcher like him deserved from this arrogant young man.

After the first few discussions, Mario began to feel that his ideas and work were as good as, if not better than, those of Dr. Levine, and he developed a keen sense of injustice that his boss was a highly paid and respected professor while he himself was merely an underpaid young researcher. One day, in the middle of a particularly heated argument, Mario told Dr. Levine outright that his thinking and research were seriously flawed. He also told his boss that he thought he could do a much better job running the laboratory than Dr. Levine did.

All hell broke loose. Dr. Levine yelled at Mario and told him to pack up and start looking for another job. Mario came back and called him incompetent. As these words came out of his mouth, however, he was suddenly intimidated by authority and feared for the future of his career. He eventually apologized to Dr. Levine. After that, he left the room and the research group with his tail between his legs. Everyone soon heard what had happened, and the specific words exchanged in the argument passed from mouth to mouth as though the conversation had been secretly recorded. Although some researchers in Dr. Levine's group shared Mario's concerns about the quality of their leader's ideas and research, they were so happy to see the young offender gone that they uniformly sided with their boss.

After leaving Dr. Levine's group, Mario worked at home, and it took him a long time to find another research job. Did he learn his lesson and swear to rethink his behavior in the future? Not really. But he conceded that his approach to the situation hadn't been well thought out. If he found himself fighting for status with a higher-ranking researcher later in his life, maybe his approach would work. But at this early stage of his career, he would have to be patient.

THE MACHIAVELLIAN STRATEGIST

One year after Gina joined Microsoft, Sarah, an experienced midlevel business manager in her late forties, was recruited to Microsoft from SunSystems. Sarah was given an office two doors away from Gina's, so Gina had an opportunity to observe how her new colleague—who had a reputation for being ambitious and successful—navigated the waters of her new working environment.

One week before Sarah began working at Microsoft, everyone in Gina's department received an email from Sarah in which she introduced herself, briefly described her background and interests, and expressed how excited she was about the prospect of working at Microsoft and, in particular, joining Gina's department. In each email, Sarah mentioned that she owned an old German shepherd named Buck. She was very attached to him, she wrote, and hoped some of her future coworkers shared her love for dogs and would be interested in walking their dogs together. In each email, there was also a line or two individually addressing the recipient: for example, Sarah told Gina that she thought they had a common interest in yoga and they should get together soon for lunch so they could talk about it. A few days after she began her new job, Sarah took all of her closest coworkers, one by one—including Gina, the employee at the bottom of the pecking order—out to lunch. During lunch, Sarah was attentive to everything the other person said. She detailed the professional interests that she had in common with her new colleague and discussed opportunities for pursuing joint projects in which they might both benefit professionally.

In reality, Sarah had no interests—personal or professional—in common with any of her new coworkers. She did have an old dog named Buck, but she paid a professional pet-sitter to walk him every day and happily left him in a kennel for days, sometimes weeks, when she traveled out of town, which she did quite often. From Sarah's perspective, the real purpose of these lunch conversations was twofold. First, she wanted to make a good first impression on everyone; she had learned from previous experience how important first impressions are and how things generally go better in the workplace if you are liked by everyone else. People are much more willing to do you favors, for example. Sarah also had a big ego—being loved by everyone was something she expected regardless of who the other people were.

The second, and more important, goal of these lunch conversations was to gather information on her new department and the people who worked in it. She wanted to find out quickly how much power and influence everyone had and identify both the dynamics of power and the key players and marginal characters: the winners and losers, who was going up and who was going down, who was friends or enemies with whom and for what reason. She also wanted to find out more about her coworkers' personalities and career trajectories, their personal and professional strengths and weaknesses, so that she could either exploit these people or protect herself from them in the future, if and when the need arose. During this information-gathering phase of her strategy, Sarah accepted every invitation she received from her coworkers, whether it was to a cocktail or dinner party or a professional group within the company. She simply wanted to show everyone what a nice person and good team player she was and get all the information she needed to prepare for the second phase of her strategy.

Sarah had transferred between companies before and knew how important this information-gathering phase was for strategic career advancement. Her goal was to become a high-ranking executive at Microsoft, but unlike Gina, she wasn't going to wait for things to happen on their own. She was going to take control and make things happen on her own terms. Although Sarah could be very self-absorbed and ruthless with others, she had learned that it's always bad to make enemies in the workplace. You never know who you might need one day, and it's always better to have the loyalty and support of everyone around you—even the people who don't count much. Sarah had learned from experience that low-status coworkers could cause a lot of damage to the career of someone they didn't like by spreading malicious gossip. And someone as socially savvy as Sarah could obtain everyone's loyalty with a relatively small investment made at the beginning. Sarah wanted to rise to the top quickly, and she knew that given her personality and her skills with people, this could be best accomplished through politics: alliance formation and social manipulation.

After about a month, Sarah had had lunch and socialized with all of her coworkers, who were already commenting among themselves about what a nice and competent person she was. What a great addition she was to the department and to the company! If only their department director could be someone like Sarah, they sighed.

At this point, Sarah switched to phase two of her strategy. After she identified a couple of key players in the department—people whose support she needed to rise quickly to the top—she began focusing her efforts on them. She quickly became indifferent to all the others, beginning with people like Gina, although she continued to smile politely at them. Sarah also quietly withdrew from all the commitments she had made to various social and professional initiatives; they were a waste of her precious time.

Sarah aggressively pursued alliances with the most powerful administrators, using any means at her disposal, from offering them professional favors to socializing, even flirting a little with one guy who seemed attracted to her. She realized that these colleagues also had some concerns about their department director; Sarah amplified these concerns by sharing some negative gossip she had heard about him at her previous company. The rest of the department didn't seem to notice the change in Sarah's behavior. She had made a good first impression, and obviously nobody expected that her lunch invitations would continue forever. It was understandable that as Sarah became busier with her new job, she would withdraw from some of the commitments she had made earlier, and soon everyone forgot about Buck the dog and his pressing need for canine companionship. In her aggressive pursuit of power, Sarah actually contributed to decisions that hurt Gina and other low-ranking employees, but they didn't notice it or simply didn't want to believe it.

In the end, Sarah's political strategies paid off. Less than one year after she joined the company, the director of her department was involved in a scandal that led to doubts about his ability to continue serving in his position. Although many people in the department continued to support him, Sarah convinced her close allies that the director was unfit for his job and had to go. At a business meeting, Sarah and her allies launched a concerted attack and eventually forced him to resign his position and quit the company. When the names of possible successors were discussed, Sarah's name was usually the first one mentioned, and she was immediately offered the position. She humbly accepted the new appointment, which came with a great deal of new decisional power and tripled her previous salary. At this point, Sarah began phase three of her strategy: she started spreading rumors about Steve Ballmer and his deficiencies as the CEO of Microsoft.

Monkey Stories

THE UNOBTRUSIVE IMMIGRANT

Billy was a four-year-old rhesus macaque male living on Cayo Santiago, a small island off the coast of Puerto Rico.[1] On this island, there is a population of about one thousand monkeys, who live in six different groups. The groups are ranked in a linear dominance hierarchy: Group Alpha, the largest group with almost three hundred individuals, is the top-ranking group, while Group Omega, which numbers only thirty-eight individuals, is the lowest-ranking. The small groups try to stay out of the way of the high-ranking ones as much as possible, but occasionally encounters do occur. These encounters between groups may result in intense battles, but they also provide opportunities for young males to check out attractive females in other groups, espy in these groups any family members who emigrated years earlier, and gather information that will help them make decisions about which group to join when the time comes for their own emigration.

Billy belonged to the fifth-ranking group, and within this group he was born into the matriline at the bottom of the group's dominance hierarchy. As he approached his fourth birthday, things were not looking good. His mother didn't want to have anything to do with him and avoided him all the time. His sisters also ignored him; they were too busy hanging out with their mother, aunts, and cousins and playing with the babies in the group. Billy had two older brothers, but both had left the group two years earlier; he had since seen them only once during a brief encounter with Group Omega.

Billy was approaching puberty, and with testosterone boiling in his veins and blurring his thinking, he couldn't help constantly staring at the females in heat. He once made a disastrous advance at an attractive young female from the top-ranking matriline in his group. As Billy walked up to her and lip-smacked the way he had seen the adult males do it (lip-smacking to a female macaque in estrus is equivalent to foreplay in humans), the female screamed her head off, with the result that her female relatives, the alpha male in the group, and a couple of his buddies all chased Billy around the island for hours. To Billy, this was a sign that it was time to go: he had to leave the group and seek his fortunes somewhere else. And so he left.

After spending a couple of weeks on his own, he started following Group Omega, where he thought his older brothers now resided. His presence at the periphery of the group didn't go unnoticed by the Omega monkeys. The alpha male and some of the other males threatened Billy and chased him away every time he came near a member of their group. Most of the adult females screamed at him, just as the females in his previous group had done. The situation didn't seem much better here. But Billy was right about his brothers being in this group—on a few occasions they approached him and groomed him for a while, which made him feel better. Moreover, an attractive subadult female in Group Omega seemed to be intrigued by Billy, and a couple of times she raised her tail, flashing her bright red behind right in Billy's face.

During all this, Billy behaved submissively to everyone in Group Omega, flashing "fear grins" left and right to any monkey within a one-mile radius. If someone approached him, he immediately withdrew and crouched or hid behind a bush. This went on for weeks until eventually the Omega monkeys tired of threatening Billy; they let him hang around while they were resting and feeding, and he traveled with them when the group moved around the island, though always hanging back.

Billy was afraid of everyone in Group Omega, including the babies, who sometimes investigated him under the vigilant gaze of their mothers. There was no question in every monkey's mind that Billy was at the bottom of the hierarchy. He had been accepted into the group, but on the lowest rung of the dominance ladder. Things stayed that way for a whole year, until a new guy, looking just as intimidated and disoriented as Billy initially did, joined the group. By this point, Billy was no longer the lowest-ranking male. Over the next four years, two alpha males left the group or died and were replaced by the males that ranked just below them, while other middle- and low-ranking males disappeared from the group for mysterious reasons. (The reasons were mysterious to the monkeys, but we know that these monkeys were captured and sold to a research laboratory.) After six years in Group Omega, Billy's situation had significantly improved. He mated every year and became the father of a few rambunctious little monkeys. He also made some friends in the group, and now, when the group traveled, Billy walked in the middle and no longer at the periphery. Thanks to the deaths or disappearances of some of the higher-ranking males and the immigration of new young and submissive males into the group every year, Billy gradually rose in rank, so that in his seventh year in the group, he found himself in an amazing position for someone who tried to avoid fights at all costs and was even afraid of thunder: Billy was in the second-highest-ranking spot. Believe it or not, he was now the beta male in the group.

Then something unexpected happened. A human researcher doing some experiments on the island captured the alpha male of Group Omega and kept him out for good. This was it, Billy's chance to become the alpha male. But unfortunately for Billy, things didn't go so smoothly. As he tried to act like an alpha male, walking around with his tail up and, for the first time in his life, threatening other females and males left and right, a pack of adult females from the dominant matriline ganged up on him and chased him around mercilessly for days. The male ranking just below Billy saw an opportunity for career advancement and joined the females in the attacks. For some reason, the females seemed to think that this guy would make a better alpha male than Billy. As a result, not only did Billy miss his chance of becoming an alpha male, he ended up being banished from Group Omega. After a year of walking around the island by himself, lonely and depressed, Billy developed pneumonia, and one day a researcher stumbled upon his dead body.

Aside from the unhappy ending, Billy's story illustrates the typical way in which rhesus macaque males leave their natal group, join a new one at the bottom of its hierarchy, and gradually rise in rank. This process has also been observed in wild Japanese and long-tail macaques—macaque species closely related to rhesus macaques— and males like Billy have been called *unobtrusive immigrants*. These male monkeys accept a "seniority" system of advancement in rank in which their status slowly rises with time spent in the group and as the higher-ranking males leave or die. This arrangement has also been called a *succession system* or *queuing system*, to convey the notion that the males patiently wait in line for their turn to become high-ranking. If they stay in a group long enough and are lucky or skilled, they may manage to make it all the way to the top. Some, like Billy, never make it to alpha status owing to either bad luck or bad manners.

For reasons I explain later, this seniority system is common in large groups of macaques. In these groups some alpha males are never challenged and maintain their status for years. Some males on Cayo Santiago have held alpha rank past the age of twenty years, even as they start to look decrepit. This wouldn't happen in the wild, where male macaques are lucky if they live past ten or twelve years. In the forests of Asia, where the macaques live, alpha males are never left to die of old age but instead are challenged in a duel—as in the Wild West of Sergio Leone and Clint Eastwood's "spaghetti western" movies—by a lone stranger who appears on his white horse out of nowhere and takes over the town by shooting the sheriff and his deputies, wasting no time in talk or politics.

The Challenger Immigrant

In 1975, primate researcher Bruce Wheatley studied the behavior of a group of long-tail macaques in a region of Borneo, Indonesia, called East Kalimantan for several months. The group Wheatley had been following included three adult males, two subadult males, and ten adult females with offspring of various ages. On March 10, he observed the following events. Three strange adult males and one subadult male entered the tree where Wheatley's macaque group was sleeping and threatened its members. One of these four intruders, named Rambo (Wheatley actually named this male GL, but Rambo sounds better), performed many branch-shaking displays in the parts of the tree where everyone could see him. The following morning the four strange males followed the troop as it foraged in the forest, and Rambo continued to act cocky and defiant, walking around with his shoulders straight and his tail up. That afternoon, when the group stopped to rest in a tree, Rambo attacked and chased the group's alpha male up and down the tree, until the alpha male eventually abandoned both the tree and the group. In merely two days, and without any bloodshed (or any help), Rambo became the group's new alpha male—the deposed king was never seen again.

Rambo was a large male, probably around seven or eight years old, in his physical prime. He had self-confidence and good fighting skills, and he was attractive to females, which explained why they readily accepted him as the new alpha male. He remained the alpha male of that group for two years, during which he mated copiously and sired many offspring. Two years later, however, a new lone stranger rode his horse into town. He challenged Rambo, as Rambo himself had done. When Rambo was seriously wounded in the fight, he was kicked out of the group. After his wounds healed, Rambo approached another group and once again challenged the alpha male. This was a group, however, of over eighty macaques, including more than ten adult males. The resident males supported the alpha male against Rambo's challenge, and Rambo got his butt kicked once again. After he wandered alone in the forest for a couple of days, one of his wounds became infected; the following day, Rambo died quietly behind a bush.

Challenger immigrants—sometimes called *bluff immigrants*—such as Rambo are males in their physical prime. They are strong and impulsive and have no patience for waiting in line. Their challenges may be successful when they join a small group, but they almost invariably fail—for reasons I will explain—when a large group is involved. In large groups, the most common way for a male immigrant to rise in rank is through succession. But not everyone is as lacking in ambition as Billy. For all the impatient ones, there is another option—illustrated by the following story of Max the challenger resident.

The Challenger Resident

Dutch primatologists Maria van Noordwijk and Carel van Schaik studied long-tail macaques in the forests of Sumatra, Indonesia, in the early 1980s. They followed several groups of these monkeys, every day, for a few years. In one of these groups, they observed an adult male they named Max successfully challenge and dethrone the alpha male.

Unlike Rambo, Max did not challenge the alpha male the day after he tried to join his group. Max had joined the group a couple of years earlier, when he was still a subadult and had not yet reached full adult size. He immigrated into the group unobtrusively, accepting a low-ranking position in the hierarchy. He had been accepted because he behaved submissively to the resident males and tried to make friends with the females, some of whom, as they got to know him better, liked him even better than the group's high-ranking males. Max kept a low profile for a couple of years, concentrating on eating good meals, sleeping well at night, and occasionally having sex with a couple of females who seemed to like him. He was careful not to do it in the presence of the alpha male, so he didn't get into too much trouble. During these two years, presumably, he carefully observed the behavior of the alpha male and the other adult males and females, creating a mental record of who was friends or enemies with whom.

One morning Max felt confident enough to launch his attack on the alpha male. The alpha male was stunned by Max's impudence, presumably became very angry, and fought back. He was helped in the process by the beta male, and their counterattack was initially successful. Max, however, did not give up his ambitions of group domination. He continued to challenge the alpha male, day after day, for two months, until finally the alpha male gave in and left the group with his tail between his legs.

As with the challenger immigrants, successful *challenger residents* are fully grown adults in their physical prime. They are generally more successful in takeovers than challenger immigrants, perhaps because of the amount of social knowledge they have been able to accrue. Having spent a year or more in the group, the challenger residents know the others and are known by them as well. Challenger residents do not attempt to challenge alpha males who have recently attained their status because they know these alpha males are strong. Knowing your opponent's strength is crucial if you want to take the risk of challenging him for the top spot. Challenger residents also probably know whose attacks can be ignored, who is likely to form strong defensive coalitions, and who should be defeated first.

Alpha males never give up their status without a serious fight, and challenges sometimes take place over the course of several weeks or even months. But because of their knowledge and strategizing, challenger residents are often successful. They launch their challenges when their chances of success are highest—for example, after the incumbent has been dominant for more than a year and has lowered his guard, making some enemies within his own group.

Models of Ladder-Climbing Strategies

The stories of Billy, Rambo, and Max show that male macaques, like new Microsoft employees or young researchers at the beginning of their academic career, have at least three different strategies to attain high rank after they immigrate into a new group. But why do these strategies exist in the first place, and how do monkeys make decisions about which strategy to use? As usual, the answer has to do with the costs and benefits. A monkey chooses a particular strategy when its benefits are greater than its costs and when the benefit-to-cost ratio of this strategy is higher than that of the others.

For the macaques, the main cost of challenging an alpha male is the risk of injury or death. To understand the benefits of being the alpha male or of being high-ranking in general, we need to understand the nature of male-male competition in the macaque world. Male macaques compete with one another to mate with females and produce offspring the way people compete with one another for money. Primatologists distinguish between two kinds of competition: contest and scramble. Imagine a group of one hundred macaques with ten adult males and fifty fertile females. One way in which the ten males can compete to impregnate the fifty females is by fighting with each other and establishing a dominance hierarchy that regulates access to the females, so that the top-ranking male mates a lot and the others mate less; the lower their rank, the less they mate. This is *contest competition.* Alternatively, each male could try to mate with and impregnate as many females as possible without directly interfering with the other males and fighting with them. This is *scramble competition.*

The extent to which male-male competition is by contest or scramble varies across different species of primates, and even across different groups of the same species. At one end of the continuum is the "super-contest." This is essentially a winner-take-all market: the alpha male mates with all the females, while the other males get little or no action. A good human example of a winner-take-all market is the Hollywood film industry, in which a select few actors and actresses receive multimillion-dollar paychecks for every film they make, while thousands of other actors are perpetually unemployed or get paid minimally for their performances. (This is also the case in the book publishing business, in which a handful of best sellers make millions of dollars, while millions of best seller–wannabe books sell only a few hundred copies or don't get published at all.) In a winner-take-all market, the odds of making it to the top are extremely low, but the benefits of being at the top are so high that many people—and many macaques—are willing to try to play the lottery.[2] All actors and actresses dream of being as successful as George Clooney and Angelina Jolie, and all male macaques dream of being alpha males.

At the other end of the competition continuum is the "super-scramble." This is a situation in which dominance rank has no effect on the ability to mate because males of all ranks have equal probabilities of impregnating females. In a group in which male-male competition is by super-scramble, dominance hierarchies don't exist because there is no clear benefit in having them (unless rank confers other advantages, like eating better food and living a longer, healthier life). An example of scramble competition in humans would be a berry-picking contest in which a group of people walk through a forest carrying baskets and each person tries to pick as many berries as possible without interfering with the other contestants.

Carel van Schaik and his colleagues developed a mathematical model that shows that, as male-male competition shifts from super-contest to super-scramble, corresponding changes occur in male strategies of rank acquisition.[3] In the model, they use a variable—represented by the Greek letter beta—to indicate the extent to which competition in a species or a group is a contest or a scramble. In the model, beta is a number between 1 and 0. When beta equals 1, we have a super-contest, and when beta is 0 we have a super-scramble. The value of beta depends mainly on how many females live in a group and whether mating takes place every month of the year or during a restricted breeding season. In the real world, beta can be estimated by examining the DNA of all the adult males and all the infants born within a group: if all the infants have the same DNA as the alpha male, it means that beta is 1, whereas if the DNA analysis shows that the infants have been fathered by many males within the group, beta is low.

When beta is close to 1—as in a winner-take-all market—there is strong pressure to use high-risk tactics of rank acquisition, because the potential benefits of alpha status are high. When beta is close to 0 and the monkeys are picking berries for their baskets, low-risk tactics like the queuing system are popular. It is important to clarify that in a winner-take-all market it is worth challenging only the alpha male, who has all the power, and no one else. This is why, when challenger immigrants join a new group, they invariably challenge the alpha male. They would not try to enter the male dominance hierarchy in the middle by challenging some of the low-ranking males. Challenging other males and risking an injury is simply not worth it because the benefits of winning are so low. Being in a winner-take-all market, however, doesn't necessarily mean that a male will challenge any alpha male he encounters any place, any time. Smart males decide to launch a challenge only when the probability of success is high.

A macaque male who wants to immigrate into a group with high contest competition must decide whether he should challenge the alpha male immediately or wait a while. Two things seem to affect this decision: the immigrating male's own resource holding potential (see Chapter 2) and the RHP of the alpha male. To succeed in attaining top rank through a challenge, a male must be in prime physical condition. The immigrant male can assess his own RHP from his age and size—whether he's big and strong or small and skinny—and also whether he feels good about himself or is low on self-confidence. Interestingly, male macaques who were born to high-ranking females in their group of origin seem to be more successful at obtaining alpha status as challenger immigrants than males who were the sons of low-ranking females in their group of origin. Although their mothers are not there to help them in their challenge, the sons of high-ranking females may be in better physical condition (because they ate more and better food growing up) and have more self-confidence than the others.

The real problem for an immigrant male considering an immediate challenge is to determine how much RHP the alpha male has—that is, whether he will be a tough opponent or a weak one. Primatologist Joe Manson has argued that an immigrant male can use the number of males in the group as an estimate of the RHP of its alpha male.[4] Manson developed a mathematical model that shows that the more males there are in a group, the more powerful the alpha male must be. Furthermore, the alpha male's RHP increases fairly slowly with an increasing number of males, but does so quickly for small groups that contain, say, between three and ten resident males. So an immigrant male macaque who understands math and is familiar with Manson's model should know that if there are only three males in a group, the alpha male is probably not that strong and an immediate challenge may be worthwhile. In a group of five or six males, the alpha male is probably stronger and has more support from the other males, so an immediate challenge may be too risky. The immigrant male may be better off waiting a few months before launching a challenge. By waiting, the immigrant male also has the opportunity to gather cues as to whether he might receive political support from other group members if he attempts a takeover (or whether they will at least refrain from joining coalitions against him).

Again, the sons of high-ranking females seem to have more success as challenger residents than the sons of low-ranking females, perhaps because they have expectations, based on their experience in their natal group, that they will receive effective support if they are involved in a fight. In large groups with more than fifteen or twenty males, the RHP of the alpha male ceases to be an issue. With so many males in the group, it is virtually impossible for an alpha male, regardless of how powerful he is, to monopolize the mating market. This means that there is scramble competition for mating—beta is low—and being the alpha male is not particularly beneficial; therefore, an immediate challenge for alpha status is simply not worth the cost. In this situation, males are expected not to take risks and to immigrate unobtrusively.

Does this mean that in large groups of macaques males have no option but to wait in line—sometimes for years—for their turn to rise to alpha rank? Not necessarily. Although males must choose between different strategies for rank acquisition at the time of immigration, there may be opportunities for advancement at any point during their residence in the new group—opportunities to cut the line and reach alpha status quickly, even overnight. While immigrating males, regardless of the chosen strategy, typically act alone, males who have been resident in a group for a long time can obtain the support of other group members, typically other males. In male macaques as in humans, a successful politician is able to form coalitions or political alliances with others. And these coalitions aren't formed only in monkey species in which males transfer between groups, such as macaques and baboons, but also in species, such as chimpanzees, in which adult males remain in their natal groups all of their lives. Chimpanzee males form political alliances with other males—sometimes with brothers, other times with unrelated males—to gain power and climb the dominance ladder in their group, to help others gain power, or to maintain the power they already have. As in human societies, struggles for power among chimpanzees rarely involve individuals acting alone; ambitious and successful individuals always operate with a strong base of political support.

Primate Politics

Before I go on about male-male coalitions in primates, it may be helpful to introduce some definitions. A *coalition* involves two or more males fighting together against a *target*, which in most cases is a single individual but occasionally is another coalition of two or more individuals. If a male is under attack, another male comes to his defense, and together they fight against the aggressor, these two males have formed a *defensive* coalition. If, instead, two or more males join forces to initiate an attack against a male who didn't previously attack either one of them, the coalition is called *offensive*.

Let's forget about defensive coalitions for the moment and focus on the offensive ones. Offensive coalitions can be formed for different reasons. As we saw in Chapter 2, when a high-ranking male baboon spends a lot of time near a female in estrus, males may pair up to form an offensive coalition against him, so that one of them might get a chance to mate with the female. In other situations, males form offensive coalitions because they want to maintain or change their dominance relationship with the target. In these cases, the fight is about power. There are three basic types of offensive coalitions: those in which the two coalitionary males are higher-ranking than the target (*conservative* coalitions); those in which one coalitionary male is higher-ranking and the other is lower-ranking than the target (*bridging* coalitions); and those in which the two coalitionary males are both lower-ranking than the target (*revolutionary* coalitions), which, unsurprisingly, turn out to be the most interesting.

Let me illustrate how revolutionary coalitions work. In June 2009, my research collaborator James Higham was studying a large group of rhesus macaques on the island of Cayo Santiago when he observed a series of revolutionary coalitions that resulted in drastic changes in the male dominance hierarchy within the group.[5] Over a period of two and a half months, a bunch of middle-ranking males repeatedly ganged up against four higher-ranking males (this was a large group comprising over twenty adult males) and eventually managed to change the dominance hierarchy and outrank the targets. Before the coalitionary attacks began, the four targets were ranked one, two, seven, and ten, so they included the alpha and beta males. The main "revolutionaries" were five males who ranked just below the tenth-ranked male, although at times they were supported by a few lower-ranking males and even some females.

The revolution began on June 1, 2009, with a coalitionary attack against the seventh-ranked male in which he sustained visible wounds. This was followed two days later by repeated attacks on the tenth-ranked male. These attacks occurred over a period of four weeks, during which the tenth-ranked male was wounded several times. On June 21, the beta male was also attacked. He soon disappeared, and Higham couldn't find him anywhere on the island. Two days later Higham finally spotted him in the sea. The aggressors had chased him into the water and continued to threaten him for hours from the beach, preventing him from swimming back ashore. Eventually, however, he made it back to the beach. The chasing and harassing of the beta male continued for about two weeks, during which time he was injured numerous times.

Finally, the alpha male was attacked, badly injured, and driven into the sea on August 10. He was kicked out of the group and never seen again. The other three targets remained in the group but dropped in rank below the tenth position, while their aggressors climbed to the top positions in the hierarchy. After looking at his field notes, Higham discovered that the revolutionary males had been in the same group for a similar amount of time and had become "friends," having spent a lot of time hanging out together and grooming one another. So one reason why these coalitionary attacks were so effective was that the aggressors "trusted" each other and worked well as a team.

The theoretical models explaining when and why males decide to form a type of coalition against other males are similar to the models that explain the different strategies for rank acquisition used by immigrant males.[6] Again, we have to consider the costs and the benefits of coalition formation. The main cost is always the risk of injury or death. The benefit of coalition formation—whether the goal is to maintain and reinforce one's status or to improve it at the expense of the target—is equal to the benefit of having high rank in general. Again, this depends on whether the males find themselves in a winner-take-all market, where the benefits of top rank are disproportionately high, or in a berry-picking contest, where rank doesn't matter. Van Schaik and other smart primatologists have developed mathematical models that explain to us common mortals why, when, and how primate males should form coalitions.

Remember that variable called beta, which tells us whether competition within a species or a group is mainly by contest or by scramble? It turns out that beta also influences *despotism*, or the steepness of the dominance ladder within a particular social system. In a highly despotic system, the ladder is set straight up and there are large gaps between the steps. This means that there are significant differences in power between top-ranking and bottom-ranking individuals, individuals don't treat those ranking lower than themselves very nicely, and climbing the ladder is difficult. In a low-despotism system, by contrast, the ladder is set on a gentle slope and the steps are close to each other, or maybe the ladder is even set flat on the ground. When there is no dominance ladder, it means that the social system is egalitarian: either all individuals have equal chances of winning fights or they don't fight much to begin with.

So, whether beta is high or low in a certain group determines whether or not there is a dominance hierarchy and how steep it is. The values of beta and the degree of despotism in a social system affect the strategies used by immigrant males to attain top rank (and also the average age of the alpha males) and the frequency of coalitionary aggression. When beta has high values—for example, in a small group where the alpha male does all the mating—alpha males are young individuals in their physical prime who get to the top by challenging and defeating the ranking alpha male. In larger groups, the values of beta decrease and alpha males are, on average, a little bit older, and in very large groups—in which beta is low—alpha males can be quite old, because they have risen to the top by succession, and this can take a long time. In despotic social systems in which beta has a high value, we would expect offensive coalitions directed at changing dominance ranks to be quite common, especially among the high-ranking and middle-ranking males. In these situations, low-ranking males are tempted to leave the group because they are excluded from mating and would be better off getting a fresh start in another group. In groups with low beta and little or no despotism, offensive coalitions should be rare or completely absent.

When offensive coalitions are frequent, it means that they are both *feasible* and *profitable*. Coalitions are feasible if they are strong enough to beat their target. They are profitable when, for each coalition member, the benefit gained in terms of increased opportunities for mating is greater than the cost entailed by the risk of injury or death. It is important to note that although risk of injury is intrinsic to any fight between two individuals, when a coalition is formed this risk becomes much higher if the coalition partner defects midfight. Therefore, coalition partners must trust each other to stay. The model developed by van Schaik and his collaborators tells us whether each type of offensive coalition—conservative, bridging, or revolutionary—is feasible and profitable and therefore whether it is frequent or rare in different social situations, depending on its effectiveness and the costs and benefits to the allies.

Conservative coalitions, which maintain the status quo, are always feasible because, by definition, the coalitionary males are higher-ranking and therefore stronger than their target. However, these coalitions are not very profitable because the coalitionary males don't gain much; they maintain the rank they already had and its associated benefits. Conservative coalitions, however, serve an important preventive function. They can be random acts of aggression to keep subordinates stressed and therefore less likely to mount challenges to dominants. Alternatively, they may serve as practice for more dangerous offensive initiatives. In other words, high-ranking males may gang up against a weak target that is unlikely to ever attack them simply to test their partners' willingness to engage in riskier coalitions of the offensive type. Conservative coalitions should be common in despotic social systems with high values of beta and are mainly formed by males near the top of the hierarchy, although not necessarily by the alpha male.

Bridging coalitions, in which one coalitionary male is higher-ranking than the target and the other is lower-ranking, are always feasible because the high-ranking member of the coalition can always beat the target on his own. However, these coalitions are not profitable for the higher-ranking coalition member unless his partner is a relative. So a typical reason for a bridging coalition would be for a high-ranking male to help a younger brother rise in rank. For example, if Billy had had the guts to challenge the male right above him in the new group's hierarchy and one of his brothers had helped him, theirs would have been a bridging coalition. Since benefits increase as the degree of despotism increases, bridging coalitions should be especially common in despotic species or in groups in which beta is high, and they should mainly involve high-ranking males, including the alpha male.

Revolutionary coalitions such as those observed by James Higham are expected in situations with intermediate values of beta and despotism, such as large groups with many adult males, because, while the cost of forming coalitions doesn't change, the profitability of revolutionary coalitions increases as despotism increases, whereas their feasibility decreases. Revolutionary coalitions can be feasible and profitable when they lead to rank improvement. These coalitions are likely to be formed by middle-ranking males—who may revolt, for example, against the alpha and beta males—because the benefits of such coalitions are highest for these individuals. This is exactly what Higham observed on Cayo Santiago. High-ranking males, however, may prevent the formation of revolutionary coalitions by preventing males from being friends—for instance, by interfering in exchanges of grooming between them. Male chimpanzees are expert at such interference.

Primate Strategies for Taking Over Microsoft Corporation

Going back to our human example, if you become a new employee of the Microsoft Corporation and want to climb the power ladder all the way to the top, it is important to have an estimate of beta—that is, how many children Steve Ballmer, the CEO of Microsoft, has fathered with the employees of his company. Of course, neither Ballmer nor any of the employees in question can be trusted with this information, so it is important to do paternity genetic analyses to establish how far and wide Ballmer has spread his DNA in the company. Depending on whether (1) Ballmer is the father of *all* the children born to Microsoft employees since he became CEO in 2000, (2) at least some of the husbands of these employees have been able to impregnate their wives or the wives of other employees, or (3) Ballmer and all the male Microsoft employees have been trying to fill their baskets with more berries than anyone else around by impregnating as many females as possible, *then* the best strategy for career advancement will be (a) challenge Ballmer in a duel right away, (b) wait a while and then challenge Ballmer, or (c) stand in line, gradually rise in rank through seniority, and patiently wait for the opportunity to become CEO without any aggressive challenges to one's superiors. Depending on the extent to which Ballmer has spread his DNA among the Microsoft children, employees can also estimate the likelihood that their revolutionary coalitions will be successful, whether they are likely to succeed in forming a bridging coalition with a lower-ranking family member or protégé, and whether their ambition to become CEO is likely to be curbed by conservative coalitions led by Ballmer and his buddies.

Well, don't take my words too literally; the parallels between human and monkey lives are not so direct. For one thing, social strategies for climbing the ladder in modern working environments have nothing to do with the number of illegitimate children fathered by the company's boss. The relevant variable in these environments is the structure of power—how despotic or egalitarian the system is, or put another way, how steep the dominance ladder is. In other primate societies, power and reproduction go hand in hand, and while this is not the case in modern human working environments, until relatively recently human society followed a similar model. In her book *Despotism and Differential Reproduction: A Darwinian View of History*, evolutionary anthropologist Laura Betzig shows that in many human societies throughout history, political and military power, which have historically been in the hands of men, had a direct link to reproduction: kings, emperors, and dictators possessed huge harems of women and sired hundreds of children with them.[7] Despots also constrained the sexual and reproductive activities of their subordinates. For example, kings in the Middle Ages demanded to spend the wedding night with each woman who married a man in their kingdom, the so-called *jus primae noctis* ("the right to the first night"). Many women conceived a child with the king on their wedding night instead of with their husband.

Regardless of whether social despotism translates directly into reproductive control, theories of primate social strategies can explain, with appropriate corrections for the species, human social and political strategies as well. First, the general principle that the viability of different strategies for power acquisition depends on the balance between the benefits and the costs of each strategy clearly has cross-species validity and applicability to human behavior. Economists know this well. More specific principles apply as well. The degree of despotism in a group or society—which is determined by the extent to which resources are monopolized and controlled by a single individual and the extent to which the despot exerts his or her power over the low-status individuals—influences the benefits of high status and the strategies to achieve it.

In a highly despotic group, particularly if the group is small and the alpha male does not have a great deal of support behind him, it pays to directly challenge the leader. Mario's strategy of directly challenging the senior professor might have been the right move under different circumstances, but in that particular situation it failed for several reasons: Mario was too young and inexperienced, and his RHP was still low relative to that of his boss. In addition, Mario failed to gain knowledge about the political strengths and weaknesses of his adversary and did not bother to get to know the other group members, much less garner their political support. Mario should have waited a few years, gained some more political experience and power, and then challenged his boss as a resident rather than as an immigrant. Military regimes are good examples of despotic systems in which the leader has a disproportionate amount of power and can be replaced only through a challenge from one of his direct subordinates—an army general or colonel—or someone from another political party, or even another country, provided the challenger has high RHP.

In more democratic human groups or societies in which power and resources are more evenly distributed among individuals on different steps of the status hierarchy, it pays to wait and rise in rank through seniority, particularly in large groups with a lot of social inertia. Gina's strategy of entering her company at the bottom of the hierarchy, being a good citizen, and keeping a low profile was appropriate given that she was young and inexperienced and therefore had low RHP and her company had a complex and multilayered structure of power. It is unlikely, however, that being patient and submissive will ever take Gina all the way to the top of the ladder. Most likely, Gina will take many years to become a midlevel administrator, and then her career will stop there. To make it to the top in a competitive environment, you must make political alliances and challenge those with power. If you don't, someone else, either a resident within the company or an immigrant from another company, like Sarah, will use more aggressive strategies and get ahead of you—exactly what happened to Billy the rhesus macaque. Sarah's strategy was the most appropriate given her own RHP—her experience, skills, and self-confidence—and the situation in the company. Sarah waited a while after immigrating into the new department and then challenged the leader as a resident, after having made the right political alliances, securing support from key allies, and even undermining the leader's power and support by spreading negative rumors about him.

In all of these situations—whether climbing the ladder as a challenger immigrant, an unobtrusive immigrant, or a challenger resident—it is fundamentally important to acquire and use social knowledge to form effective political alliances. Humans are political animals, but their societies and relationships are more complicated than those of other primates. As discussed in Chapter 2, social knowledge, political alliances, and dominance status are intimately interconnected. Social skills are necessary to form strong relationships and to be liked by others enough for them to be willing to cooperate. The ability to charm and lead others—what we call charisma—is an important skill in forming effective political alliances. And in any kind of human social organization, from academic departments to business companies to entire countries, strong political alliances are necessary for anyone to climb to the top of the ladder and stay there for a while. Clearly, having high RHP is a prerequisite to becoming a leader in every kind of human social organization. But in humans, even more than in other primates, good social skills, the ability to form political alliances, and the self-confidence that usually accompanies these traits are much more important components of RHP than the strength of one's biceps or the sharpness of one's canine teeth.

Chapter 5

Cooperate in the Spotlight, Compete in the Dark

Coffee, Tea, and Human Nature

My research laboratory at the University of Chicago is in the Biopsychological Sciences Building. In one corner of the building's lobby is a small kitchen with a sink, a refrigerator, and a microwave oven. Tucked behind the refrigerator is a large, ultramodern espresso machine that resembles a laser printer more than a beverage dispenser. The espresso lovers in my building buy massive amounts of expensive coffee once every few months, and the coffee maker/laser printer is always fully loaded and ready to go. You see, caffeine is the fuel that keeps the brains of many researchers running, and some of us like to take highly concentrated doses—in the form of double or triple espresso shots. Without drugging our brains with caffeine, it would be impossible to stare at a computer screen for hours without falling asleep.

On the wall above the coffee maker is a list of all the names of the Espresso Club members. Each time club members make themselves a cup of coffee, they mark an X next to their name. At the end of the month the marks are counted and people pay their tab to the club's treasurer. My colleagues and fellow espresso drinkers happen to be honest people who record every cup they drink, but if they wanted to cheat, there's nothing to prevent them from doing so. In theory, we could have installed a webcam near the espresso machine to monitor whether people honestly record their coffee consumption, but we didn't. We have other, more important things to worry about, such as dwindling funding for research.

(Incidentally, when webcams were first invented, one of their first applications was thought up by a guy who used it to check his coffee on the kitchen stove while he was working on his computer; the website showing his coffee maker on the stove became very popular, and people were amazed at this new technological development, which used a personal computer and the Internet to keep an eye on the room next door.)

The arrangement that my colleagues and I have developed to pay for espresso is not unusual. Researchers and students in the Department of Psychology at Newcastle University in England put coins in an "honesty box" every time they make themselves a cup of tea or coffee or use milk.[1] They have the option of putting in the box the exact amount of change for their drink (fifty British pence for coffee, thirty for tea, and ten for milk), more if they feel especially magnanimous, or nothing at all if they are out of change or out of generosity. A notice with the instructions for payment is displayed on the wall above the counter where the honesty box and the coffee- and tea-making equipment are located.

In 2006, Melissa Bateson and Daniel Nettle—two researchers at Newcastle—published, along with one other colleague, the results of an ingenious experiment involving their coworkers and the honesty box. Next to the instructions for payment they placed a banner that alternated each week between an image of a pair of eyes and an image of flowers. The eyes varied in sex and in head orientation, but all were chosen such that they were pointing directly at the person making the drink. The tea and coffee drinkers who served as guinea pigs in this experiment had no idea why the banner was up or why the images were switched every week; they probably thought it was some kind of game made up by colleagues with too much time on their hands. Bateson and her colleagues made sure that supplies of tea, coffee, and milk were maintained to keep up with demand, and every week they measured the volume of milk consumed as a proxy for total

Figure 5.1. *Cues of Being Watched Enhance Cooperation in a Real-World Setting. Photo by Dr. Melissa Bateson. Modified after Figure 1 in Bateson, Nettle, and Roberts (2006). Reprinted with permission of the Royal Society of London.*

beverage consumption. Then they calculated the ratio of money collected in the box to the volume of milk consumed in that week to take into account the fact that people didn't always drink the same amount of coffee or tea with milk every week. The researchers discovered that the amount of money people put in the box fluctuated from week to week, as the images alternated. People paid nearly three times as much for their drinks when eyes were displayed on the banner—the eyes on the wall seemed to exert an automatic and unconscious effect on the drinkers' perception that they were being watched while they paid for a drink, which made them more generous.

The honesty box experiment at Newcastle University was actually inspired by a study conducted a few years earlier by two researchers at the University of California–Los Angeles named Kevin Haley and Daniel Fessler.[2] Haley and Fessler asked over two hundred undergraduate students to come to their research laboratory and play a computerized economics game known as the Dictator Game. In this game, player 1 receives a particular amount of money and is told that he can give player 2 any fraction of this amount. Player 1 has full control over the decision to share the money (hence the name), and player 2 has no active influence on the decision. It's not a real game, like the Prisoner's Dilemma, but simply a decision-making test that taps into people's generosity.

You might think that selfish people would keep all the money and give nothing to the other player, but in reality most people playing this game choose to share some of their money. Haley and Fessler paired each student with another student and randomly designated them as either player 1 or player 2. Player 1 received $10 and had twenty seconds to give player 2 part of it—any amount between $0 and $10—while keeping the remainder. The players were told that no one would know the decisions they made and that the game would be played only once. Neither student could see or communicate with the other, so the game was played under conditions of strict anonymity. Unbeknownst to the students, the desktop background of the computers used during the experiment was being manipulated by the researchers: half the time the players used a computer whose desktop background showed two stylized eyelike shapes (eyespots), and half the time they used a computer in which the word CASSEL was displayed across the same portion of the screen, using the same color scheme as the eyespots.

It turned out that the students who served as player 1 donated 55 percent more money to the other player when the desktop background had the eyespots (on average, they donated $3.79 of their $10) than in the other condition (in which they donated, on average, $2.45). As with the experiment at Newcastle University, the authors of this study suggested that the unconscious perception of being

Figure 5.2. Nobody's Watching? Subtle Cues Affect Generosity in an Anonymous Economic Game. Photo by Drs. Kevin Haley and Daniel Fessler. Modified after Figure 1 in Haley and Fessler (2005). Reprinted with permission of Elsevier.

watched made people more generous with their money. In the last few years, many more studies have been conducted to examine the effects of "eyespots" on honesty, generosity, and cooperation. With a few exceptions, the studies have confirmed the results obtained by Haley and Fessler and by Bateson and her collaborators.

So why is it that photos of eyes—or even stylized images of eyes—result in the perception of being watched? Can't people tell that they're just pictures and that no one is really watching? Of course they can—if they think about it consciously. But if they barely notice the eyes, the images automatically activate brain responses that unconsciously influence their decisions. Many animals, including humans, are instinctively attuned to eyes and eye gaze direction.[3] For some animals, detecting the eyes of a predator means the difference between life and death. Fish, for example, are more likely to flee from objects that resemble the eyes of a predator than from similar objects that don't. Some butterflies have large spots on their wings so that if they are about to be eaten by a bird, they can spread their wings and startle the bird with what looks like the eyes of a large animal. For highly social animals, such as domestic dogs, monkeys, and apes, detecting the eyes and assessing the eye gaze direction of a dominant individual can play an important role in avoiding aggression. Dogs avoid stealing forbidden food if they see that a person is watching them, but do it promptly if the person has her eyes closed, has her back turned, or is otherwise not looking at the dog. When rhesus macaques detect the gaze of a higher-ranking group member or a human on them, they bare their teeth and rapidly smack their lips to signal submission. Low-ranking chimpanzees avoid taking food they want if they can see another individual's face and eyes.

Humans, of course—being both highly visual and highly social creatures—are more attuned to eyes and eye gaze direction than all other animals put together. Human babies who are a few days old prefer to look at drawings of faces—even if these faces only have eyes—than at any other images. In our everyday social lives—say, when we ride in an elevator with another person—we continuously monitor the gaze direction of other people and process information about their identity, facial expression, and even emotions and motivation. It turns out that in both monkeys and people certain brain cells are automatically activated by the perception of eyes and their orientation or by objects that resemble eyes. For example, some cells show spikes of electrical activity in response to a picture of an individual with opened eyes. Some cells in the brain of a macaque respond even more strongly when the monkey is looking at a picture of another monkey gazing at the camera directly than when the monkey in the picture is looking somewhere else. Most of these cells are located in areas of the brain called the inferior fusiform gyrus, the superior temporal sulcus, and the amygdala. Brain activation in response to eyes and eye gaze is largely automatic and involuntary. This explains how even the image of eyes on a wall or on the desktop background of a computer can lead to an unconscious perception of being watched.

And why does the perception of being watched make people more generous or less likely to cheat? A quick answer to this question is that the areas of the brain that process the perception of being watched are closely interconnected with other brain areas that are involved in making decisions—like decisions about whether to be honest or to cheat, or about how much money to donate to someone else.

So upon seeing eyes on the walls or on our computer desktop, the being-watched parts of our brains send electrical signals to the decision-making parts.

The long answer will take up the rest of this chapter. In the first half, I show that when people know or think that their identity is known, they tend to help others and act in ways that establish or strengthen a reputation for being generous, cooperative, or trustworthy. In the second half, I show that when people know or think that their identity is anonymous, they become more competitive than cooperative and tend to hurt rather than help others.

As in the previous cases, evolutionary biology and economics help explain why we behave the way we do. Social influences on decision-making processes, however, are something that economists have only recently incorporated into their models. Economists used to think that people always make rational choices that maximize their gains, and that they make choices in isolation from their social context, without regard for the consequences of their behavior. Researchers studying animal and human social behavior from an evolutionary perspective, however, have discovered that maximizing one's fitness—like maximizing one's earnings—often depends on taking others into account. As we'll see, the integration of economic models with evolutionary theory and the findings of animal and human behavioral research can result in more sophisticated and more predictive models of human decision-making and ultimately help bridge the gap that still separates economic and biological explanations of behavior.

The Altruist in the Spotlight

How to Choose Trustworthy Partners for Cooperation

In the last chapter, I mentioned that in both nonhuman primates and humans issues of trust are very important when choosing allies. Imagine a middle-ranking macaque male who solicits the help of the alpha male in a coalitionary attack against the beta male in the group (an example of a *bridging* coalition). After weeks of planning and scheming, the middle-ranking male finally launches his attack on the beta male, and the alpha male initially helps him. If, however, the alpha male changes his mind midfight and switches sides, the consequences of his defection for the middle-ranking male are disastrous. The beta male will naturally want to retaliate, and with the alpha male now on his side, he will make dog meat of the erstwhile challenger. So, whether you are a macaque or a human, choosing a trustworthy coalition partner is crucial for the success of any political enterprise. The question is how to find one.

Our family members generally make good coalition partners—think about how the Bush family members have helped one another's political careers—because we have shared genetic interests and we can estimate, knowing how they behaved in the past, whether or not they can be trusted in the future. But how can we know whether an unrelated individual we've just met is trustworthy? To answer this question we need to revisit our discussion of cooperation and the Prisoner's Dilemma in Chapter 2.

If we play a single round of the Prisoner's Dilemma with a stranger, knowing whether the stranger has previously cooperated or defected with others can influence our move. Remember that, in the absence of any information, it's always safer to defect, but if we have information that leads us to think that the other player will cooperate, then it's better to cooperate because both players can then win big. So when trying to establish whether a potential cooperation partner is trustworthy, his or her reputation can make a big difference. Our own reputation, obviously, is equally important to the other player. Anyone with ambitions for a political career—which, by definition, requires cooperation with other individuals—should make every possible effort to establish a good reputation before running for office.

In the Prisoner's Dilemma, two people accused of committing a crime can both avoid jail time if they cooperate and tell a consistent story when interrogated separately by the police. The importance of reputation for cooperation has also been demonstrated with other economic games—such as the Dictator Game used by Haley and Fessler—in which people need to make decisions about sharing money with another person.[4] A player is more likely to be generous if he or she is told that the other player has a reputation for generosity. A player is also more likely to donate money to another player when by doing so he or she can establish or strengthen a reputation for generosity, which may be useful for future cooperation. In fact, studies have shown that in both the Prisoner's Dilemma and the Dictator Game players are more likely to cooperate or be generous if they know each other's identity than if identities are kept anonymous. As we'll see later, having a good or bad reputation can play a major role in all human affairs that require cooperation, from joint business enterprises to politics to romantic relationships.

It turns out that reputation matters in animal affairs as well. Take the example of cooperation between the large coral reef "client" fish and the tiny "cleaner" fish that swim inside their mouths to clean them. To evolutionary biologists who study animal behavior, this is a textbook example of mutualism—an interaction that benefits both parties. The large fish get their teeth and gums brushed without paying exorbitant dentist's fees, and the little cleaner fish get free food without being eaten by their clients. As is always the case when there is cooperation between two parties, however, there is also the potential for conflict of interest, which may lead to cheating. For example, the cleaner fish could cheat by biting pieces of healthy tissue off the walls of the client's mouth instead of just eating the dead tissue and parasites, and the clients could cheat by closing their mouths and swallowing the cleaners as soon as they are done with their oral hygiene session. According to University of Cambridge biologist Redouan Bshary—now at the University of Neuchatel in Switzerland—cleaners do sometimes cheat, and this becomes clear when the client makes short jolts in response to being bitten. It's sort of like saying, "Ouch!" when your dental hygienist pokes your gums—unintentionally, I hope—with one of her sharp instruments.

Bshary showed that clients choose cooperative rather than cheating cleaners, not on the basis of personal experience, but on reputation.[5] They observe cleaners at work on other clients and make a mental note of whether the clients jolt from a bite. When the time comes, they then approach cleaner fish that behave themselves and avoid cleaners that they have observed cheating. Cleaners that co-operate or cheat, therefore, acquire a good or a bad reputation, re-spectively, among their clients. According to Bshary, however, the cheating cleaners find ways to cheat about their reputation and there-fore eventually beat the system. They behave themselves in the mouths of small client fish that don't have fleshy mouth tissue, only to cheat the larger clients after being chosen by them as oral hygien-ists. In the jargon of evolutionary biologists, cheating cleaners coop-erate in low-payoff interactions to build a reputation that allows them to exploit other individuals in high-payoff interactions. I discuss the cleaner fish and their clients, and more generally the issue of finding good partners for cooperative enterprises, in more detail in Chapter 8.

WHY IT's So DIFFICULT TO PROTECT THE ENVIRONMENT

As nicely illustrated by the honesty box experiment conducted at Newcastle University, in everyday social situations people's ten-dency to contribute to a common enterprise, such as the shared cost of purchasing coffee for the espresso machine, may be influ-enced by events or factors that they perceive as improving or dam-aging their reputation—such as being watched while being generous or cheating. Although the issues involved in deciding to cooperate or defect in the Prisoner's Dilemma and to make or not make an hon-est contribution to the Espresso Club are similar, there is an important difference between these two scenarios. In the former, the dilemma lies in the potential conflict between the interests of the two prisoners. As we saw earlier, this model applies to all situations in which co-operation is required between two individuals, such as in the forma-tion of an alliance against a third party or the mutualistic interaction between a client fish and a cleaner fish. The dilemma involved in deciding whether or not to contribute money to the Espresso Club, however, lies in the conflict between the interests of an individual and those of the group. Obviously, the group does best when all play-ers cooperate and everybody pays for their coffee. However, econo-mists and evolutionary biologists tell us that, unless the individuals who don't cooperate are identified and punished, nobody should ever contribute anything to a public enterprise. To prove that their game theory models are correct and their cynicism is justified, they point to the results of experiments conducted with so-called public goods games.

The classic public goods game consists of four or five players who are given the options of investing a token in either a private good—for example, keeping the token for themselves and saving its entire value—or a public good such as contributing it to a public pool. After contributions are made, the content of the pool is doubled—this provides an incentive for cooperation—and is then divided by the number of players and evenly paid out, irrespective of each player's contribution. If all of the players contribute their tokens to the public pool, they all win and make the maximum possible return from their investment. However, if one or more players behave selfishly and keep the token for themselves (in the language of game theory these individuals are referred to as *free riders*), then contributing to the pool is no longer recommended for anyone. In fact, mathematician John Forbes Nash—the man with a *Beautiful Mind*, played by actor Russell Crowe in the film—developed mathematical models showing that each token paid into the pool yields a return of only a half-unit to the contributor and therefore nobody should ever contribute anything to the pool.

Consistent with the predictions of the models, experiments have shown that when multiple rounds of the public goods game are played, players initially cooperate but then gradually become more selfish as they are increasingly tempted to cheat and take advantage of the system. People's failure to contribute to public goods explains why managing a limited public resource, such as pollution-free air or water, is so difficult. An American ecologist named Garrett Hardin aptly termed this phenomenon "the tragedy of the commons" in an influential article published in the journal *Science* in 1968.[6] The tragedy is that, when it comes to protecting the environment and other similar matters, the individual's interests prevail over the public ones and in the end everyone loses.

It's the same with paying taxes to the government or to the state. We should all spontaneously pay taxes because it's in our own interest to do so: taxpayer money is used to build roads and public schools, to fund scientific and medical research, and in many countries—which thankfully will soon include also the United States—to provide universal health care coverage. If tax-paying is not enforced, however, and tax evaders are not punished, very few people will do it. In the United States, everyone has heard scary stories about being audited by the IRS and having to pay hefty fines or go to jail for failing to pay taxes. People are willing to try to evade taxes anyway—particularly if they make a lot of money and the stakes are high—but not nearly as many as in Italy, where tax payment enforcement is lax and no one is scared of getting caught and being punished for tax evasion. As a result, tax evasion is rampant in Italy, most Italians are wealthy because they keep their tokens for themselves, the country is always bankrupt, and in the end that hurts everyone, just as a degraded environment does.

There are many "tragedies of the commons" in human societies, but again, these situations are not unique to our species. For example, think of a situation in which several different parasite microorganisms grow in the same host—the "public resource." The interest of each individual parasite would be to exploit the host as much as possible; if all parasites behave this way, however, the public resource is over-exploited, the host dies, and the parasites die along with it (or at least have to find a new host).

The tragedy of the commons can be resolved not only by enforcing cooperation and punishing free riders—as is done with tax payment and tax evasion—but also by providing rewards to the individuals who spontaneously cooperate. A good reputation, for instance, can offset the cost of cooperation. To better understand why and how reputation can influence people's tendency to cooperate in public goods situations, it is necessary to introduce the notion of indirect reciprocity. Evolutionary biologists distinguish between direct and indirect reciprocity. *Direct reciprocity* refers to a situation in which an individual altruistically helps another with the expectation that the beneficiary will later reciprocate. If reciprocation in fact occurs, both individuals benefit. In *indirect reciprocity*, an individual altruistically helps another, but the help is returned by a third individual, not by the original recipient of the help. Typically, all three individuals involved belong to the same group and have some interests in common, so that if indirect reciprocity becomes a common practice within the group, then all group members will benefit. Indirect reciprocity, however, can also work when help is offered to individuals outside one's group (for example, by making a donation to a charity organization) because the donor gains—not directly through support received by a third party, but through an enhancement of his or her reputation. Thus, by helping others who are unable to reciprocate the help to the donor in the future, people build up a good reputation—or, in the jargon of game theorists, a "positive image score"—whereas refusing to help can damage their reputation.[7]

Having a good or a bad reputation for generosity can mean the difference between good and bad business, or between political success and failure. For example, when Bill Gates was still the CEO of Microsoft and before he married his wife Melinda, he rarely if ever made large donations to charities, despite being often at the top of *Forbes* magazine's list of the richest men on earth. His reputation for being stingy probably didn't help Microsoft's business, although the software giant was so powerful and successful that, in the end, the reputational damage was largely inconsequential. After Bill married Melinda and their Bill and Melinda Gates Foundation started making large donations to charities, Bill's reputation and Microsoft's image improved considerably, and this probably had positive consequences for the business as well.

In case you don't find the Bill Gates example compelling, consider the experimental studies that have demonstrated how concerns about reputation can influence people's willingness to contribute to public goods and how a good reputation can translate into monetary or political gains. A few years ago, two economists, James Andreoni at the University of Wisconsin and Ragan Petrie at Georgia State University, had two hundred college students play a computerized public goods game in conditions that varied in degrees of anonymity. The students played in groups of five, and each player in a group was given twenty tokens, which could be invested in a private good that paid the investor two cents for each token invested or in a public good that paid the investor one cent for every token invested. There were four experimental conditions. In the *baseline* condition, all five players in a group knew the total contributions of their group to public goods, but didn't know who the other group members were or how much they had individually contributed. In the *information* condition, the five players knew exactly what each group member had contributed to the public good, but didn't know who they were. In the *photos* condition, subjects saw photos of other group members, but had no information on their individual contributions. Finally, in the *information-and-photos* condition, subjects saw photos and received information, so they knew who their group members were and how much each had contributed. The experiment showed that providing information and identification together (the information-and-photos condition) resulted in a 59 percent increase in giving to the public good relative to the baseline condition. Two other economists, Mari Rege and Kjetil Telle, later replicated these results in a study in which Norwegian students living in Oslo who had never met before played a single round of a public goods game. Thus, even strangers can be induced to invest more in public goods by concerns about reputation.[8]

The effect of the perception of being watched on cooperation that Haley and Fessler demonstrated in the dyadic Dictator Game has also been demonstrated in public goods games. Harvard researchers Terence Burnham and Brian Hare had students play multiple rounds of a public goods game under conditions of anonymity; however, in one condition the students used a computer that displayed on the screen an image of Kismet, a robot built at the Massachusetts Institute of Technology that doesn't look particularly human—except for its eyes. It turned out that players who perceived that they were being watched by Kismet contributed 29 percent more to the public good than did the players with a neutral computer screen.[9]

Social psychologists believe that we care so much about reputation because we all continually seek approval and respect from others to maintain our self-esteem or to promote our social identity. In their

Figure 5.3. *Kismet, the robot built at MIT. Photo from Wikipedia.*

view, the psychological reward of establishing a good reputation explains why people are willing to invest in it. This may well be true, but there may be deeper, more selfish incentives at work. Economists argue that humans invest in reputation in order to maximize their personal financial gains and minimize their losses, while evolutionary biologists think that animals do it in order to maximize their personal fitness gains and minimize their losses. (Here "financial" means "money" and "fitness" means "survival and reproduction.") Essentially, a good reputation is an extension of our spending limit on our credit card. With no reputation, we get no credit from others. As we build a reputation with acts of cooperation, there is a corresponding increase in others' willingness to give us credit for larger and larger amounts in future business transactions.

That a good reputation can result in financial or political gains has been demonstrated experimentally. For example, a series of studies by a Swiss biologist, Manfred Milinski, and his colleagues showed that reputation enhances cooperation in public goods games when these games are alternated with indirect reciprocity games in which players are indirectly rewarded for their generosity. Thus, somewhat like the cheating cleaner fish that gain a good reputation in one context and then benefit from it in another one, good reputation built through generosity in the public goods games results in gains in the subsequent indirect reciprocity games. However, as Milinski showed, if the indirect reciprocity games are eliminated, or if the players have different identities in the two types of games so that the reputation built in one game cannot be transferred to the other, then contributions to the public goods quickly disappear.[10] Thus, people are aware of whether they will be recognized in a future social situation and use this information to invest in their reputation only if doing so is likely to result in tangible gains in the other context.

This sounds pretty cynical, but again, it's what the experiments show. One may be a little skeptical of these results, however, because they are all obtained from college students in artificial experimental conditions. Maybe people in the real world don't act like college students earning a few bucks by volunteering to serve as subjects for an economics experiment.

In the real world, as it turns out, revealing the identity and generosity of givers is important. Charities, for instance, often give their donors considerable opportunities to be identified, from building statues in their honor to publishing their names in a magazine or on a website. In addition, by offering donors premium gifts that vary in relation to the contribution amount, fund-raisers allow donors to broadcast to others that they gave a certain amount. Comparisons between levels of donations are important for reputation and status-building among donors. Fund-raisers seem to know this: when they solicit contributions from a particular donor, they may disclose what others have already given and suggest a contribution amount that would allow the donor to be competitive with his or her peers. Charities also promote comparisons in generosity among donors—and therefore competition for reputation—by reporting gifts in categories. For example, museums and theaters list donors in their programs by categories such as "patrons," "sponsors," and "fellows," based on the amount of their donations. Carefully constructed by the institution, these categorizations are most likely intended to get people to "round up" their donations to get into a higher category.

Andreoni and Petrie have provided us with further experimental evidence that reputation plays an important role in donations. The two economists had students play a computerized game that mimicked charity donations. Players were given the option of remaining anonymous, but when they chose to have their identities revealed, they were assigned to different categories of generosity based on the size of their gifts. People contributed more when they were given the option of having their contribution announced. Category reporting also had a significant effect of shifting gifts up to meet the lower bound of the higher category. Similarly, another experiment by Milinski and collaborators showed that donations made in public to UNICEF, a well-known world relief organization, resulted in both personal financial gain (the player-donors received more money from the members of their group) and enhanced political reputation (they were elected to represent the interests of their group).[11] As evolutionary biologist Richard Alexander eloquently wrote in his book *The Biology of Moral Systems*, "In complex social systems with much reciprocity, being judged as attractive for reciprocal interactions may become an essential ingredient for success."[12]

In personal relationships, a person's reputation is often based on direct observation of his or her behavior in previous interactions, while in public life reputation can be formed through well-advertised acts of cooperation or generosity. In both cases, transmission of reputation through third parties—gossip, in other words—can also be crucial. We all know that gossip can play a huge role in establishing or destroying a reputation, so the notion that gossip can influence one's tendency to cooperate or be generous should not be surprising. Indeed, there are experiments that remove all doubt on this point. In a recent study conducted by psychologists Jared Piazza and Jesse Bering, people played the Dictator Game anonymously, but some player 1s were told that the player 2s would discuss their decision to share money with a third party who knew their identity. The threat of gossip and concern about their reputation prompted the dictators to be more generous in sharing their money.[13]

Given how widespread gossip is in human societies—studies investigating the content of conversation in college cafeterias and in tribal villages have reported that over 50 percent of it consists of gossip—it is clear that anyone who has business or political ambitions should make an effort to be the subject of positive rather than negative gossip.[14] Cultivating a good reputation, however, can be an expensive investment, and doing so only makes sense if there is a good chance that it will translate into future gains. The game theory models predict that people will stop investing in building their own reputation as soon as they find out that no future gain is likely to result, and the experiments show that this is indeed the case.

The Punishment of Defectors

Building a reputation through flamboyant acts of generosity, such as making a million-dollar donation, is not an option for most of us. More generally, securing a good reputation is not enough of an incentive for many people to invest in public goods. For these folks, contributions to public goods must be enforced by laws and threats of fines or jail time. No external factor, however, is as powerful (and as cost-effective) as the internal control that people can exert on their own behavior. When rules are internalized—that is, when people feel a sense of ownership and believe that obeying the rules is in their best interest—then they become the most efficient enforcers. People simultaneously function as their own informants, police officers, and judges to make sure they catch themselves if they break a rule and give themselves the appropriate punishment. Self-punishment can be harsh and painful—consider the practice of self-flagellation among some medieval Catholic monks, who whipped their own backs to punish themselves for impure thoughts or actions (a phenomenon brought to our attention by the albino monk assassin in *The Da Vinci Code*). People can also give themselves the death penalty by committing suicide out of guilt for something they did.

Laws that force people to pay taxes are usually not internalized—no one commits suicide because they cheated the IRS—but religious and moral rules are. This explains why religion and morality are far more effective means of controlling people's behavior than the laws and law enforcement agencies of democratic societies, or even the violence and intimidation of oppressive dictatorial regimes. Some people are better than others at internalizing rules, or find it easier to do it in some contexts than in others. When feelings of guilt don't work to restrain people's selfish behaviors and tendencies to cheat, then others lend a hand with what evolutionary biologist Robert Trivers calls "moralistic aggression."[15]

When people are caught defecting in cooperative interactions—whether they involve another individual or a whole group—others will punish them by condemning their behavior in public or by spreading negative gossip. By giving defectors a bad reputation, others inflict costs on them by undermining their viability as partners in future cooperation—whether in romantic or marital relationships, business partnerships, or political activity. That people are willing to go out of their way to punish defectors has been demonstrated by countless experiments involving the Prisoner's Dilemma, the Dictator Game, public goods games, and other economic games involving cooperation and trust.[16] But let's turn away from the experiments to take a look at more concrete examples from everyday life.

We are all aware of the damage that malicious gossip can do to someone's social, financial, or political reputation. Malicious gossip is a form of punishment that allows the moralistic police to do potentially lethal damage to defectors without exposing themselves to the risk of retaliation. In some cases, the defector being punished is not even aware of the gossip. When things at work or at home take a bad turn for the defector, he or she may just blame it on bad luck or karma. Other forms of moralistic punishment—the most interesting—are better advertised.

The purpose of moralistic aggression is to inform everyone that cheating has been noticed and is disapproved. A mild form of this is horn-honking. For example, I have the aggressive driving style typical of many Italians. When I drive on the streets of California and fail to respect some traffic rules, other drivers honk at me even though they are not directly affected by my driving. But moralistic aggression against traffic rule transgressors such as myself is nothing compared to moralistic aggression against people who cheat on their spouses or in sports, business, or politics. To give just one example, some wives who have been cheated on by their husband have paid exorbitant sums of money to put their cheating husband's name or face on huge posters displayed in busy urban areas to give him a bad reputation and make sure no other women will pair up with him in the future.

Punishment of individuals who fail to cooperate has been shown to exist in many animal societies in which cooperation is important. For example, rhesus macaques normally give calls to alert other group members when they discover a tree full of ripe fruit in the forest. Primatologist Marc Hauser reported that some macaques eat all of the fruit themselves without alerting the rest of the group, but that they are later attacked by the group if they get caught.[17] I wouldn't call this moralistic aggression—there is no morality among macaques— but the circumstances are quite similar to those of human moralistic aggression.

People are afraid of moralistic aggression, and rightly so: the costs inflicted by others for failing to cooperate and for breaking the rules can be high. There are therefore two reasons why we are more likely to cooperate if we perceive that we are being watched or our identity is known: in addition to building a good reputation for cooperation, which might bring tangible benefits from future investors, we also want to avoid being punished if we get caught cheating. Cheating someone who expects us to cooperate is always selfish and may also be immoral or unlawful. Defecting in a cooperative game—regardless of the nature of the game—also signifies something else: competition. When we cheat a partner or a group, we want our interests to prevail over theirs. We make a conscious choice to compete instead of co-operating. Just as cooperation has benefits—such as the brownie points in reputation we accumulate if our altruism is well advertised— so competition has costs, and these costs can be minimized or avoided altogether if our competitive/selfish behavior is hidden under the blanket of anonymity. Just as people who choose to cooperate with others like to be in the spotlight—so that everyone can see, appreciate, and eventually, they hope, reward their acts—when people choose to defect and hurt rather than help others, they prefer to operate in darkness.

Competing in the Dark

THE NEW YORK CITY BLACKOUT OF 1977

On July 13, 1977, at about 8:30 P.M., damage to power lines and relay stations caused by multiple lightning strikes plunged a large part of New York City into a complete electricity blackout that would last for about twenty-four hours.[18] When the lights went out and millions of people were left in the dark, a crime wave the size of a tsunami hit the city, particularly in its poorest neighborhoods. With the darkness guaranteeing anonymity and hindering police intervention, people looted stores, burglarized apartments, smashed windows, and set fire to entire blocks of buildings. In one night, over fifteen hundred stores were looted throughout the city. Theft and property damage were also accompanied by violent crimes: people were robbed and shot on the streets and in their homes, women were raped, and over five hundred police officers were injured. By the end of the following day, when electricity was finally restored, over four thousand people had been arrested—the largest mass arrest in the history of New York City. The number of people who committed crimes that night but didn't get caught, however, was probably much larger.

Clearly, not all of the people who looted stores or robbed, raped, or killed someone that night were career criminals. Many of them probably had no criminal record. There is an Italian proverb, *L'occasione fa l'uomo ladro*, that means, "Opportunity turns man into a thief." This proverb implies that the world is not divided into bad people who steal and good people who don't, but rather that, given the right circumstances, anybody can turn into a thief—even a murderer. According to another Italian proverb (*I proverbi sono la saggezza dei popoli*), "proverbs are people's collective wisdom." They reveal some fundamental truth about human nature, but usually come with no explanation attached. To explain why "opportunity turns man into a thief," we need rational models of behavior.

The economic and evolutionary biological models tell us that stealing is basically a selfish act of competition in which individuals benefit at the expense of others. In a "good" opportunity, the benefits are high and the costs are low. To make stealing costly, society has laws that protect people and their property and punishment for those who break them. Obeying these laws, like the laws that impose taxes on all citizens, is equivalent to being forced to contribute one's tokens to the public pool in a public goods game. There is a price to pay for not contributing—or worse, for stealing tokens from the public pool. When the price is eliminated, the benefit of cheating is no longer offset by a cost.

The anonymity and immunity afforded by darkness prompt people to break the social contract and unleash their selfish and competitive tendencies at the expense of others—both individuals and society as a whole. Usually, the people who do so are the ones who stand to benefit the most—the poor and the oppressed. (Millionaires don't need to loot an appliance store to obtain a new TV set.) These are the people who feel that in the cooperative game enforced by the social contract they get the short end of the stick.

Competition, whether regulated by rules such as in sports or by laws against crime, is as integral a part of human nature as cooperation, and fluctuations in the ratio between the benefits and costs of competition can unleash harmful behavior that is normally restrained. Since it's mostly poor and uneducated people who sit in jails, does that mean that the human biological propensity to defect in a cooperation game when it's advantageous to do so is stronger in these people? Do education, wealth, and job stability protect us from our potentially harmful competitive tendencies and make us more likely to play by the rules even when it's advantageous to break them? I don't think so. The well educated and the wealthy have the same tendency as everyone else to defect under the right circumstances, but this tendency is expressed in different contexts. To give an example, being in the dark and protected by anonymity unleashes the harmful competitive tendencies of one particularly well-educated group of people: university professors.

ANONYMOUS PEER REVIEW

Many important decisions concerning the political, legal, and economic life of a country, as well as the health and well-being of its citizens, are influenced by intellectual advances made in disciplines such as political science and law, economics and sociology, biology and medicine. Progress in these disciplines, in turn, depends on funding to conduct research and on the publication of research findings in journals and books. Grants and publications also benefit the universities in which the research is conducted. In the United States, when professors obtain a large research grant from the government, more than half of it goes directly into their university's pockets. In the United Kingdom, the more professors publish articles in prestigious journals, the more their university is funded by the government. Not surprisingly, professors who are highly successful in securing funding and in publishing are sought after by universities and rewarded with quick promotions and high salaries. The pursuit of knowledge, once the occupation of scholars writing books with pen and paper at home, has now become big business, and many large universities increasingly operate like corporations.

Needless to say, given how much is at stake, publishing articles and obtaining research grants have become very competitive activities; only a small fraction of articles submitted to journals are accepted for publication, and an even smaller fraction of grant applications are funded. Who decides which are accepted? Politicians? Special experts hired by the government? No, it's the professors themselves. Using a process called *peer review*, they read each other's work and recommend acceptance or rejection. Thus, when a reviewer recommends acceptance, he or she is not only promoting a worthy enterprise that can help society and improve the future of humankind, but is also boosting the author's career and inflating his or her bank account. Conversely, the consequences of having a paper or a grant application rejected can be devastating for a researcher, both psychologically and financially. As in any human affair in which people negotiate with one another the outcome of their endeavors, personal interests figure prominently in the review of intellectual and scholarly work, and human nature can get in the way of making objective judgments and decisions.

Whoever invented peer review—I'll call this person "the Inventor"—must have realized that every time an article or a grant application is reviewed, the reviewer's recommendation to accept or reject the author's work is equivalent to a cooperating or defecting move in a standard Prisoner's Dilemma Game. If the professors review each other's work multiple times over the course of their careers, the game is played repeatedly with frequent role reversals. The reviewer makes the first move with his or her recommendation to accept or reject the author's work. If the author plays tit-for-tat, as expected, he or she will match the reviewer's move the next time they interact and their roles are reversed.

The Inventor must have realized that if professors played these games with each other, decisions made in the peer review process would simply track the behavior of the players, regardless of merit. This would be a disaster; mediocre individuals could have spectacular academic careers, and a great deal of taxpayer money could be wasted on not-so-serious research projects. Worse than that, asymmetries in the frequency with which the two players take the reviewer or the author role (for example, a senior professor might review the work of a young colleague much more often than the other way around), or in the consequences of their recommendations (for example, a single recommendation by a reviewer might affect an author's entire career), might lead to attempts to influence decisions. An author might try to encourage or reward a positive recommendation through bribery, or use intimidation or violence to discourage or punish a negative recommendation. And university professors do occasionally resort to violence to punish peers for their rejection: not too long ago, a biology professor in Alabama whose tenure application was rejected by her colleagues shot and killed three of them.[19]

Protecting the reviewers and keeping them safe from the wrath of rejected authors was a primary concern of the Inventor, who thought that this could be effectively accomplished by keeping the reviewer's identity anonymous. This way, the reviewer wouldn't have to worry about playing tit-for-tat and would be immune to bribery attempts and fear of punishment. The Inventor assumed that once the reviewer was freed from all external influences, he or she would make honest and objective recommendations about the author's work because it would be the right thing to do (and because he or she had signed an agreement to that effect). Being so concerned about how human nature would influence the author's behavior, the Inventor overlooked the fact that reviewers are human beings too, and that any human being who is given the power to make decisions about another's work, career, and financial success under conditions of anonymity is very tempted to use this power to his or her own advantage.

The problem is that, although professors are more qualified than anyone else to review each other's work, they also belong to the same community and compete for limited resources such as grants, publications, and high-status positions. Altruistic motives to advance the goals of the community—centered on producing high-quality and important work—inevitably mix with selfish motives to advance their personal agenda at the expense of competitors. Models from economics and evolutionary biology show us that the relative prominence of these altruistic and selfish motives depends, again, on the balance of the benefits and costs of cooperation and competition. Anonymity dramatically alters this cost-benefit balance in favor of competition, because anonymity reduces the benefits one can obtain through cooperation (reducing incentives to cooperate) and virtually eliminates the costs of competition (providing incentives to engage in selfish behavior).

To understand this better, recall that people are more inclined to cooperate with others and to contribute to their community if doing so enhances their good reputation and increases their prospects for future personal gains. In the peer review system, objective and honest reviews of other people's work benefit the community as a whole, but anonymity deprives the reviewer of the opportunity to gain a good reputation through cooperative behavior. In addition to reducing the indirect benefits of cooperation, reviewer anonymity also greatly reduces the costs of competition. Economic and evolutionary models predict that if individuals are given the opportunity to hurt their competitors without paying a price—without being punished for it—they are likely to do so. People's behavior during the 1977 New York City blackout is consistent with these models. The same happens during wartime when soldiers invade another country and commit crimes against civilians, or following natural disasters such as earthquakes or major hurricanes, which cause a breakdown of the law enforcement system with a consequent spike in crimes. Is the behavior of anonymous reviewers also consistent with these models? When darkness falls, do university professors murder their colleagues and loot their property?

OPPORTUNITY TURNS THE REVIEWER INTO A COMPETITOR

Investigating the effects of reviewer anonymity on the peer review process gives us a window into human nature. Among other things, university professors do a great deal of research on their own behavior, and the peer review system has been the subject of hundreds of studies. Sadly, the results of these studies and anecdotal evidence from our colleagues tell us that anonymous reviewers sometimes do indeed loot the intellectual property of the authors whose work they review (when they steal ideas from these authors and delay the publication of their work to give themselves time to redo it and claim it as their own) and also sometimes damage or destroy the reviewed authors' property for good (when they suppress the publication of their articles or the funding of their grant applications with harsh negative criticism). In addition to property theft and property damage, anonymous reviewers can also be guilty of crimes against the person—which may amount to professional "murder"—such as when they recommend that a colleague's application for tenure be rejected.

A few years ago I requested funds to conduct a particular research project by submitting a grant application to the National Science

Foundation (NSF), a U.S. government agency that funds a lot of scientific research. When the NSF receives a grant application from a professor, it asks professors in the same field at other universities to provide anonymous reviews and recommend acceptance or rejection. The reviewers are carefully instructed by the NSF to comment on the scientific merit of the proposal and its potential impact on society at large without making any personal comments on the author or the author's work in general. This recommendation notwithstanding, one anonymous review of my application began with a statement along these lines: "The author [that would be me] received a lot of funding from the National Institutes of Health [another government funding agency] in the past, and if he thinks he is going to get money from the NSF too, he is wrong." So much for limiting comments to the proposal's scientific merit.

Receiving an anonymous review with undeservedly harsh criticism and personal attacks (we call them *ad hominem*, from Latin) is a psychologically traumatic experience that hurts a researcher beyond the professional damage caused by the rejection itself. To survive the anonymous peer review system and make it in academia, a researcher must develop a thick skin and the ability to shrug off rejection and the career setbacks that come along with it. I have been rejected hundreds of times in my career and like to think that my skin has grown thick in the process, but receiving harsh anonymous reviews still makes me want to quit my job and take up gardening instead. Plants can be much safer and more reassuring to be around than people.

Clearly, not everyone uses the anonymous peer review system to shoot down their competitors, or at least, not all the time. Many rejections are well deserved, provide constructive criticism, and help the author learn how to produce better-quality work. More often, the anonymous reviews of an article or grant application are a mixed bag, containing both good ones and bad ones. A study conducted by two British doctors, Peter Rothwell and Christopher Martyn, and published in 2000 in the journal *Brain* showed that agreement between independent reviewers as to whether manuscripts should be accepted, revised, or rejected was not significantly greater than that expected by chance.[20] If a manuscript of mine receives three anonymous reviews, it will often be the case that one says the article is fabulous, one says it's just okay, and the third says it's the worst piece of junk ever written. (Note: This mainly happens to *good* articles; *bad* articles usually get three negative reviews.) At some journals, a manuscript with mixed reviews still has a chance at publication, whereas at journals that reject most of their manuscripts a strong negative review is the kiss of death, especially if it comes from an influential senior professor, as is often the case.

Trying to publish papers and obtain grants through the anonymous peer review system is like walking through a minefield: the land mines are everywhere, and each step you take is bound to set one off. Most of the land mines are small, however, and the explosions don't cause deadly injuries. You'll be able to resume the march after getting back on your feet and licking your wounds. However, the continuous blows, along with the anxiety, fear, and anger that follow each explosion, can take a psychological toll and impose a significant cost in time and resources as you are forced to proceed slowly, zigzagging left and right, and even to take a few steps backwards more often than not.

Every now and then, you encounter people in academia who have tried to beat the system by establishing a safe corridor through the minefield that allows them to proceed straight, at a steady speed, avoiding the stress and pain of the continuous explosions. Some of these folks have published hundreds of articles in their long careers, but most of those articles have appeared in just one or two journals, where they never receive a rejection by virtue of their personal connections with the journal's editors. I've watched some of these folks become overconfident in their success and decide to step out of their safe corridor and submit a manuscript to a different journal. Invariably, they hit a land mine and are blown up just like everyone else.

Vampires Versus Werewolves

It should be clear by now that there is a great deal of subjectivity in the peer review process; reviewer anonymity allows people with an ax to grind to shoot down their competitors with impunity. But human beings don't compete only on a one-on-one basis. They also belong to groups and compete to advance the interests of their group over those of other groups. In the business of research and academia, competition may involve people from different countries and pits men against women, old against young, professors from large research universities against those from small colleges, researchers doing human research against those who use animals, and researchers who study monkeys against those who work with laboratory rats. So, when a reviewer and an author belong to different groups, the anonymous review provides an opportunity for the reviewer to score points against a competitor group. I am a Monkey-Man, and when I submit a grant application for peer review, I am terrified that it might fall into the hands of the Rat-People. They want to exterminate *all* of us, regardless of who we are (because our animals are cooler than theirs), and they want all of the funding for animal research to go to *them*. It's like the Vampires and Werewolves in the Twilight series. That these struggles happen in real life and not just in the minds of paranoid professors has been shown by many studies of the peer review system.

Some of the more interesting studies have compared the outcomes of manuscripts reviewed with single-blind peer review and those reviewed with double-blind peer review. Single-blind review is the traditional system in which the reviewer is anonymous but the author isn't, while in the double-blind review the author is anonymous too. To make the author anonymous, the first page of a manuscript or grant application, which contains the author's name and personal information, is removed before the document is sent to the reviewers. It turns out that reviewers are more likely to recommend rejection when authors are women, citizens of another country, or professors belonging to competing institutions to a much greater extent in single-blind reviews than in double-blind reviews. In other words, when reviewers know the author's identity, their reviews express all kinds of biases against particular groups of people. (Note: Double-blind review is better than single-blind review but far from perfect: in many cases reviewers can still guess who the authors are.) The bias against women is particularly serious because it may contribute to forcing women out of research and academia. Although an increasing number of women obtain PhDs and begin academic careers, many more of them are likely to drop out of academia than men.[21]

Age cohort effects have not, to my knowledge, been formally studied with comparisons of single-blind and double-blind peer reviews, but it is well known in academia that members of the same age cohort regularly help each other's careers—whether because of past ties or common interests—at the expense of other age cohorts. For example, the members of the baby boom generation (born between 1945 and 1960) are a powerful group in American academia: many of them hold positions of power that allow them to make life-or-death decisions about the careers and success of younger researchers. (I discussed the nepotistic behavior of baby boomers in Chapter 3.) Instead of helping the younger generations, many baby boomers hold on to their power and resources with all the means at their disposal, crushing the professional hopes and ambitions of the generations that follow them. I know of senior professors who have obtained funding from the U.S. government for thirty years in a row and who consistently write harsh anonymous reviews to thwart the attempts of young scientists to obtain their first grant. As a government grant administrator who oversees hundreds of grant reviews once said to me, "Senior scientists kill their young." (Unless the young are biological or adopted relatives, of course.)

Studies of the peer review system have also shown that when authors request that particular individuals be excluded as possible reviewers because of competition, their manuscripts and grants are more likely to get accepted than when the reviewers are chosen without the author's input. Again, this shows that reviewer competition is a real phenomenon. There are also studies showing that when reviewers sign their reviews and reveal their identity, their reviews are more likely to contain constructive, rather than destructive, criticism. Finally, harsh anonymous reviews of articles submitted to prestigious journals with a rejection rate higher than 90 percent are especially common. Some authors submit only their best manuscripts to prestigious journals and end up receiving—even as they report their newest and most exciting findings—the worst reviews of their lifetime.

If you are a scientist who has received an unfair rejection from a prestigious journal such as *Science* or *Nature*, you are in very good company. Juan Miguel Campanario, a physicist at the University of Alcalá in Madrid, Spain, has compiled an online archive of more than thirty instances in which an article describing an important discovery in science or medicine that later resulted in its author being awarded the Nobel Prize was initially "trashed" by an anonymous referee reviewing it for a prestigious journal.[22] Campanario has also documented instances in which articles reporting a Nobel Prize–worthy discovery were immediately and harshly criticized upon publication by fellow researchers who wrote letters and commentaries to dismiss the new discovery. Campanario's collection of rejections and criticisms experienced by Nobel Prize winners neatly illustrates the hurdles that professors have to overcome to publish their best work and get it recognized by their peers.

The examples in Campanario's online archive are just the tip of the iceberg. The history of every discipline is full of examples of articles reporting important discoveries that were either harshly rejected initially or ignored for a long time. The truth is that outstanding papers are more difficult to publish than papers of average quality, just as it's more difficult to obtain funding for original and innovative research projects than for conservative projects that entail only minor extensions of research already done before. There are many possible explanations for this phenomenon: papers with claims of important discovery are scrutinized more carefully than others; many such claims turn out to be unfounded and therefore rejection is warranted; new ideas are difficult to understand and don't fit into existing paradigms; or the whole process of scientific progress is very conservative and proceeds with small steps. All of these explanations are partly correct, but I think competition between researchers plays an important role as well, and reviewer anonymity makes competition a lot more prominent.

Some of my colleagues think that my views of the peer review system are overly pessimistic. They concede that anonymous reviewers are sometimes sloppy or wrong and may hurt authors unfairly, but this outcome is rarely intentional, they argue, and has little to do with competition. By and large, the anonymous peer review system has a lot of support, which explains why it's still so popular. My view—that reviewers should reveal their identity in order to take responsibility for what they write or pay a price when they abuse the system—is the minority view and always faces strong resistance. The main objection is a human nature argument: everyone is afraid that authors would retaliate against reviewers who reject their work— even if the rejection is fair and justified. Supporters of the anonymous peer review system are certain that reviewers are always honest and professional, but they seem to have very little confidence that authors would play by the rules, not take rejection personally, and restrain their impulses to seek revenge against unfavorable reviewers.

This raises an important question: why is it that the influence of human nature on the author's behavior is so readily recognized while the influence of human nature on the reviewer's behavior is so easily dismissed? Here is a possible explanation. The author's retaliation against an unfavorable reviewer can be interpreted as a form of self-defense, and we all accept that humans have a strong instinct for self-defense. The notion that reviewers could take advantage of anonymity to hurt their competitors, however, implies that we have an instinct for offense, for unprovoked aggression, and that we express it when anonymity shifts the costs and benefits of competition and makes it advantageous. Although the tendency for unprovoked offense is as hardwired in our brains as self-defense and retaliation, self-defense is easier to justify on moral (and legal) grounds. We expect that people will try to hurt those who have hurt them (even the Bible recommends "an eye for an eye, a tooth for a tooth"), but hurting others who haven't done anything, merely because it's possible and personally advantageous to do so, is considered morally reprehensible. Hence, to recognize that reviewers may use anonymity to hurt their competitors is to acknowledge that human beings are morally corrupt (or simply amoral) creatures. It's better to think that there are a few bad apples out there who can be neutralized.

Although "people's collective wisdom" and the rational models of human behavior developed by economists and evolutionary biologists tell us that "opportunity turns man into a thief," many find this view of human nature hard to swallow. It's a lot more comforting to think that humans are guided by moral principles and religious beliefs than to admit that cost-benefit ratios usually hold sway. Rational explanations of human behavior are labeled cynical because they leave no room for morality or religion. Even the most modern, civilized, and religious societies, however, force their citizens to cooperate with one another and to curb selfish competitive tendencies by manipulating the costs and benefits of cooperation and competition— for example, through the law enforcement system. The police are out on the streets every day to make sure we don't give in to the temptation to steal from and murder others and to ensure that we are swiftly caught and punished if we do. Many, however, prefer to think that the world is divided into good and bad people and that the police exist to protect the good folks from the bad ones.

Well-educated people such as university professors are—like everyone else—sensitive to the changes in the cost-benefit ratio of their behavior brought about by anonymity, and reviewers can abuse anonymous peer review for competitive purposes despite signing agreements in which they promise not to. Human beings are more likely to behave themselves and to be generous toward others when they are being watched—or when they think that they are being watched—because they expect to be rewarded when they help others and punished when they hurt them. When darkness falls and anonymity reigns, however, all bets are off.

Even in the darkness, of course, some people choose to behave themselves and to help others. If I find myself in Central Park in the middle of the night during the next power outage in New York City, I certainly hope that everyone around me will choose to be a good citizen. For all those who are in their homes when the lights go out, I'd recommend staying inside and locking the door. At all other times, while you are conducting social transactions that require cooperation, my advice is this: keep the lights on and let others know you're watching.

Chapter 6

The Economics and Evolutionary Biology of Love

What Went Wrong in Beverly Hills?

If you believe the tabloids, film stars Jennifer Aniston and Brad Pitt met in 1998 on a blind date set up by their agents. Both were beautiful, successful, and ready to settle down in a stable relationship. Less than two years later, they were married in a blowout wedding in Malibu. After a nine-month search for their perfect love nest, they bought a $13.5 million, 12,000-square-foot mansion in Beverly Hills, which they completely renovated over a two-year period. Included in the renovation was the addition of a nursery, as it was clear that the couple had plans for starting a family. Jennifer and Brad also became business partners, cofounding the production company Plan B Entertainment, which produced several successful films, including *Troy* and the updated *Charlie and the Chocolate Factory*. The couple enjoyed a long (by Hollywood standards) period of marital bliss, during which they were featured on the cover of *People* or *US Weekly* more frequently as a happy couple than in the magazines' usual reports of marital infidelities, spousal battery, or other celebrity drama. In November 2003, however, Brad met Angelina Jolie on the set of their new film, *Mr. and Mrs. Smith*, and the two fell in love. Brad and Jennifer announced their separation in January 2005, and their divorce was finalized in October 2005, by which point Angelina was already pregnant with Brad's child.

What went wrong? Was Brad and Jennifer's love not real? Was their commitment not strong enough? Did they not want the same things from life? Did they discover things about each other that made them change their minds about being together? If Brad wasn't sure that Jennifer was the perfect partner for him and was looking for an Angelina type, why did he decide to settle down with Jennifer?

All of these questions have already been answered a million times by tabloid reporters, celebrity biographers, psychologists, psychiatrists, astrologers, and hundreds of other relationship and celebrity experts. As far as I know, however, no tabloid reporter ever interviewed an economist or an evolutionary biologist about the Brad and Jennifer marriage fiasco. And, you might ask, why would they? What do economists and evolutionary biologists know about love and relationships? Well, more than you might think. As it happens, economists and evolutionary biologists have a lot of the answers as to why marriages and romantic relationships work or don't work. They will even go so far as to define what love is and why it exists in the first place.

The Economics of Love

Two Economists' Views

According to the University of Chicago economist Gary Becker, author of *A Treatise on the Family*, we choose mates who best promote our material interests and then remain in the relationship as long as the benefits outweigh the costs.[1] When the costs go up and the benefits go down, we call it quits. From Becker's point of view, there is nothing mysterious about what happened to Brad and Jennifer. When they first met, they could give each other what they wanted, and they continued to do so for a few years, so they both benefited from being together. Then the circumstances changed; for one or both, the benefits of being in the relationship weren't enough, while the costs started rising (like the cost of missing opportunities to date other people). When the costs outweighed the benefits, the relationship ended. This analysis exemplifies one way in which economists think about romantic relationships: we make rational decisions to start them and then to end them, and love is merely an afterthought. But not all economists agree.

Economist and public intellectual Robert Frank lies more on the romantic end of the spectrum. In his 1988 book *Passions Within Reason: The Strategic Role of the Emotions*, he rejects coldhearted cost-benefit analyses of romantic relationships and says that love is important, even as he tries to explain its existence with economic arguments.[2] His view of love is embedded within a more general theory of the origins of emotions that blends biology and economics to try to explain why we have feelings and how our feelings help us deal with some of the problems of our everyday lives.

Like Becker and other economists, Frank views romantic relationships as cooperative ventures in which two individuals choose to stay together to pursue joint goals—such as raising children or accumulating property (or producing movies, if you are Hollywood actors). Unlike the other economists, however, Frank is an optimist—he doesn't believe that people always pursue selfish interests and that their behavior is necessarily the result of rational choice. The hallmark of romantic relationships, he maintains, is that two partners make a commitment to each other and stick together—or at least they try to—even when the cost-benefit ratio becomes unfavorable to one partner or both. How is that possible? As we saw in previous chapters, cooperative relationships between unrelated individuals can be a tricky business. Economists tell us that long-term cooperative relationships pose what they call a commitment problem.

To illustrate this problem, Frank uses the following example. Imagine two guys—Smith and Jones—who want to open a restaurant together. Their complementary talents and skills make their cooperation advantageous: Smith is a talented cook, Jones is a good manager. If they each worked alone, their potential would be quite limited. Working together, however, each partner has some opportunities to cheat. Smith can take kickbacks from food suppliers, while Jones can steal from the cash drawer. If one of them cheats, the other one will lose big, while if they both cheat, they both stand to lose a great deal. It's a classic Prisoner's Dilemma. Playing tit-for-tat in this situation, however, is not an option—the first time either of them gets caught cheating, that's the end of their restaurant. So rather than constantly checking each other's moves and living with the anxiety of being cheated, Smith and Jones choose a different strategy: they make a commitment never to cheat each other and sign a contract to that effect.

If Smith and Jones were purely rational beings and made decisions based on the relative costs and benefits of cooperating versus cheating, their commitment would mean nothing and would be doomed to failure. The problem is that no matter how beneficial it is to both partners to cooperate when they first start their joint business, sooner or later the circumstances will change and it will become advantageous to one partner, maybe to both, to cheat. It becomes very difficult to resist the temptation to cheat, especially if getting caught is unlikely. The partnership may stand a chance if the future circumstances that favor cheating can be predicted in advance (for instance, if opportunities for cheating predictably arise every ten years, Smith and Jones could simply sign a ten-year contract), or if Smith and Jones behave irrationally and are willing to pass up opportunities for cheating when they present themselves. Unfortunately, Smith and Jones cannot predict when future opportunities for cheating will arise, nor do they know each other well enough to be able to predict each other's behavior when circumstances change.

Even though the dilemmas of cooperation and the strategies to solve them are similar in animals and humans, as we saw in the previous chapter, humans have come up with creative ways to make an individual stick to a commitment to cooperate even when it would be more advantageous to defect. First, there are reputation effects: when other people (future business partners) find out who cooperates and who cheats, there is a boost in reputation that comes from cooperating, and a cost that comes from cheating. Second, there can be sanctions on cheating that make it unappealing regardless of the rise of favorable circumstances. Finally, there are internal mechanisms of control involving morality and feelings, which other animals presumably don't have. Smith and Jones may stick to their commitment never to cheat each other simply because they think it's wrong to cheat and would feel guilty if they did. If people who sign a contract of collaboration make a moral commitment to each other and back it up with the appropriate feelings, then there is a chance that the commitment might actually last.

In a nutshell, Frank's theory of feelings is that they exist to help people solve the commitment problem. His idea is that although human beings presumably evolved as selfish creatures whose behavior was entirely controlled by cost-benefit ratios, human social lives have become so complex that people need to rely on long-term cooperative relationships with unrelated individuals to survive and flourish. In these long-term cooperative relationships, it is imperative that we curb our selfish impulses and ignore the temptation to cheat even if it's advantageous. Whether our sense of right and wrong is a "biological instinct" evolved by natural selection or the internalization of a social contract provided to us by our parents, society, and culture, morality is effective in making people cooperate regardless of their best interests. Feelings can help in this regard. Not only do we think it's wrong to break a commitment, but we also feel guilty about it and regret the pain we have caused to the other party. These negative feelings can be a strong deterrent to breaking a commitment—though not to sociopaths who don't feel anything—but positive feelings are important too. We feel good when a commitment is first made, and feel even better if it's maintained for a long time, possibly forever. And here's where romantic love enters the picture.

Love: The Perfect Business Solution?

From an economist's point of view, all cooperative relationships are business partnerships: whether the goal is to make children or to run the Planet Hollywood restaurant chain (a joint business venture launched in 1991 by actors Arnold Schwarzenegger, Sylvester Stallone, and Bruce Willis) makes no difference. If these relationships are to last long enough for their goals to be accomplished, they will present the usual commitment problem. And as usual, the problem is solved by a combination of reputation effects, imposed sanctions, morality, and feelings. In the case of romantic relationships, before the commitment problem even arises, the two business partners must find each other. Robert Frank helps us understand how this process works, using another business analogy.

The search for the perfect romantic partner shares many features with the search for the perfect apartment in a rental housing market, Frank says, or if you are a landlord, the search for the perfect tenant. It takes time and effort to search for and inspect available rental apartments, just as it takes time and effort for the landlord to interview potential tenants and assess their reliability. Waiting until you have inspected all the apartments or interviewed all the potential tenants before you make a final decision would mean never making a decision. There are too many empty apartments and potential tenants to begin with, and every day new apartments show up on the market and new potential tenants call landlords. Instead, if you're seeking an apartment you visit a few apartments, and if you're a landlord you interview a few potential tenants to get a sense of who is out there. Then, according to Frank, when both parties meet a sensible quality threshold (that is, they find something that's good enough), they terminate their search and decide to settle. At this point, the commitment problem arises, and the two parties try to solve it by signing a lease.

Signing the lease is necessary for two reasons. The first is that when an apartment-seeker finds a landlord, or a landlord finds a tenant, who meets their quality threshold, they never have the required information about this person's past or the ability to predict his or her future behavior that would be necessary to make a good choice. The tenant might pay rent on time for a few months and then start skipping payments. Or the landlord could be helpful at the beginning and then refuse to make necessary repairs in the apartment.

Acquiring all the necessary information about the two individuals to predict their future behavior would take forever. If a lease is not signed, the partnership between the tenant and the landlord will deteriorate as soon as either party does something wrong or unpleasant.

The analogies with the restaurant partnership are obvious. Even if the tenant and landlord each behave perfectly and provide no reason for breaking their partnership, neither party can ever be sure of having found the best possible deal; in theory, the tenant could find an even better apartment a month later, and the landlord could find someone willing to pay more the next day. If both continue the search for a better deal, it is guaranteed that sooner or later their partnership will end. To make it work, they must end their search.

Without a lease, the business partnership between landlord and tenant would be unsustainable. The uncertainty of being in a partnership with someone who might end it at any time is stressful. Moreover, if either partner could end the relationship whenever the benefits of leaving were high enough, the eventual breakup could be very costly to both partners. Both parties, then, are better off restricting their potential for rational choice and limiting their options. By signing a lease, each party guarantees loyalty to the other no matter what and gives up any opportunity to accept a better deal that might become available during the period covered by the lease. This way, both parties gain by the stability of their situation and avoidance of the unforeseen costs that could arise from its breakup.

Like apartment-seekers and landlords, people want stable and long-term relationships but have limited time and opportunities to find partners. They start sampling potential partners, and when they find someone who meets their quality threshold, they decide to settle down. Once a partner is chosen, however, circumstances often change. People discover new aspects of their partner's personality or behavior that they didn't notice before, or their partner's behavior changes, or another, more attractive person comes along. One way or another, sooner or later, opportunities for cheating or breaking up arise. Given the large investment made in the relationship, this would be very costly and potentially disastrous for the joint goals of the relationships. To minimize the chances of this happening, people sign a marriage contract that imposes significant financial penalties on the partner who breaks up the partnership, such as hefty lawyer fees, alimony payments, and child support (for further information about this, please ask Tiger Woods). If the cause of the breakup is cheating with another person, there are also potentially high reputation costs—such as having one's name printed on a giant billboard and being subjected to public moral condemnation (see Chapter 5). But all of these deterrents may not be enough to prevent a relationship breakup.

The circumstances could change dramatically so that the costs of staying together become very high, or the benefits drop, or Angelina Jolie walks in one day and all bets are off. In that case, financial penalties won't matter, damage to reputation won't matter, and morality and feelings of guilt and empathy for the other's pain won't matter. Something else is needed, an irrational force that doesn't act as a deterrent against breaking up but rather makes people want to be together no matter what the circumstances are, no matter how bad the cost-benefit ratios, no matter what other people think, and no matter how emotionally devastated Jennifer Aniston is. The irrational force is love; love trumps reason, money, reputation, morality, and empathy. According to Frank, love is the ultimate solution to the commitment problem, the only one that can ensure that two people stay together. He argues that relationships motivated by irrational love are more successful than those motivated by material self-interest or exchange and cooperation. "Do people in love relationships really set aside material self-interest?" Frank asks. His answer: "There is evidence that many do."

In *Passions Within Reason*, Frank launches a crusade against his fellow economists to prove that rational models of human behavior are inadequate and that people often behave against their selfish interests. "As the rationalists emphasize," Frank writes,

> we live in a material world and, in the long run, behaviors most conducive to material success should dominate. Again and again, however, we have seen that the most adaptive behaviors will not spring directly from the quest for material advantage. Because of important commitment and implementation problems, that quest

will often prove self-defeating. In order to do well, we must sometimes stop caring about doing the best we can.[3]

The existence of love as a solution to the commitment problem in romantic relationships is the ultimate demonstration that "to do well" we have to behave irrationally, ignore the costs and benefits of our decisions, and embrace pure altruism and its costs.

Is Frank right? Does love really exist as a solution to the commitment problem? Let's examine Frank's ideas a little more critically. I can think of at least three specific problems, and a more general one, with his ideas.

The first problem is this: if romantic relationships present the same commitment problem as any other cooperative business partnership—as Frank implies—why is it that business partners don't fall in love to solve their commitment problems? If signing an apartment lease is as ineffective as signing a marriage contract in maintaining a partnership, why is it that tenants don't fall in love with their landlords all the time? Since love occurs in (some) romantic relationships but not in any other type of human cooperative partnership, there are two possible conclusions: either love is a solution to the commitment problem but both the problem and the solution are different in romantic and business partnerships (that is, romantic partnerships present unique problems that require unique solutions), or love is not the solution to the commitment problem in romantic relationships.

Second, is it really true, as Frank maintains, that a relationship motivated by irrational feelings is inherently more stable than one motivated by rational thinking and prospects for material exchange? In fact, one could argue just the opposite—that the irrationality of love makes the romantic relationship subject to mercurial whims, while a partnership established on the basis of rational reasons is likely to persist as long as the reasons do. I don't know if Bill and Hillary Clinton still feel passionate love for each other, but it's clear that they have both benefited greatly, professionally and financially, from their partnership, and their relationship seems very stable.[4] All those who expected them to divorce after the Monica Lewinsky scandal and the end of the Clinton presidency were proven wrong. According to the tabloids, Brad and Angelina are very much in love, but they seem to fight all the time and are always on the verge of breaking up. Maybe what keeps them together is not really their love but all the interests they share, including children and joint properties.

If it's true that love provides a solution to the commitment problem, how long does the solution last? This is the third problem with the commitment model: it does not explain when, how, and why love between two people ends. If love is an irrational feeling that occurs independent of the costs and benefits of being in a relationship, it follows that changes in these costs and benefits would not cause love to end. Although true love lasts forever, the kind of love available to common mortals seems to be strongest at the beginning of a relationship, when passion is at its peak, and to gradually fade until, in some cases, it simply disappears. The commitment model, however, would predict the opposite temporal pattern. According to this model, when two people start a relationship, their partnership is advantageous to both: they have common interests and want to pursue common goals because going alone is either impossible or less effective than a joint venture. In short, love is not really necessary at the beginning. According to the commitment model, love is needed later, when circumstances change and it's no longer advantageous for one or both partners to be together. So love should get stronger and stronger with time, to make sure that these irrational feelings hold the relationship together when rational arguments would push for a breakup. Is this what really happens?

There is another, more general issue with the commitment theory of love: it does not account for the notion that romantic love is not only about the maintenance of closeness and commitment to a partner but also often about the pursuit of a relationship (sometimes unrequited) with an object of desire. To illustrate this point, I turn to one of the masterpieces of European literature: the 1897 play *Cyrano de Bergerac* by Edmond Rostand.[5] Here is a brief plot summary:

In Paris, in the year 1640, a brilliant poet and swordsman named Cyrano de Bergerac falls deeply in love with his beautiful, intellectual cousin Roxane. Cyrano, unfortunately, has a large ugly nose and does not want to tell Roxane how he feels for fear of being rejected.

Instead, Roxane tells Cyrano that she loves one of his cadets, the young and handsome (but intellectually challenged) Christian, and asks Cyrano to protect him. Cyrano goes even further and begins to write love letters to Roxane on Christian's behalf. Roxane soon falls in love with the author of the letters, whom she assumes is Christian. One night Christian stands in front of Roxane's balcony and speaks to her while Cyrano stands under the balcony whispering to Christian what to say. Frustrated by Christian's romantic incompetence, Cyrano eventually shoves him aside and, under cover of darkness, pretends to be Christian and woos Roxane himself.

Roxane and Christian get married, but soon after both Christian and Cyrano are sent to the front lines of the war with Spain. During the long war, Cyrano writes to Roxane every single day, using Christian's name, and risks his life each morning by sneaking through the Spanish lines to a place where he can post the letters. When Roxane comes to the front lines to see Christian, Christian forces Cyrano to tell her the truth, but just as Cyrano is about to do so Christian is shot and killed, so Cyrano cannot tell Roxane the truth.

Fifteen years later, Roxane lives in a convent and Cyrano visits her every week. One day Cyrano, who has made many enemies in his life, is ambushed and hit in the head. He appears at the convent, walking slowly and with a pained expression on his face, but sounding as cheerful as ever. As night falls, Cyrano asks to read Christian's last letter to her. He reads it, and when it is completely dark he continues to read, as if he knows the letter by heart. Roxane realizes that Cyrano wrote the letters—she has found the soul she was in love with all along. Cyrano removes his hat, revealing his wound. Roxane exclaims that she loves him and that he cannot die. But Cyrano cannot survive his wound, and he collapses and dies, smiling as Roxane bends over and kisses him.

If neither Becker's nor Frank's economic models can explain Cyrano de Bergerac's feelings and behavior, then what can? Are there other theories of love that can help us understand it?

Perhaps biology can provide the answers.

The Evolutionary Biology of Love

SEX, LOVE, AND HAMMERS

If you ask a random person on the street what love is, there is a good chance this person will say something about finding the perfect partner, being sexually attracted, developing passionate feelings, and spending the rest of one's life with this person. An evolutionary biologist who happens to walk by and overhears this conversation, however, might object that love has little to do with sexual attraction or partner choice. From an evolutionary standpoint, sexual attraction exists to facilitate sexual intercourse, and sexual intercourse exists to make reproduction possible. Biologists and evolutionary psychologists know a great deal about how sexual attraction works and what different people find attractive and why.[6] Although sexual attraction and romantic love sometimes go together, this is not necessarily the case. Similarly, although you might think that both sexual attraction and romantic love influence, or even determine, partner choice, in reality the choice of a long-term relationship partner can be entirely independent of sexual attraction and love. Why people pair up with particular individuals and not others is complex, and understanding it requires a combination of theories and knowledge produced by many different disciplines, including anthropology, biology, economics, psychology, and sociology. But understanding why people fall in love with each other does not require knowing anything about sexual attraction or partner choice.

The fact that love exists in all human cultures and that poems and songs at every stage of human history have described it in similar terms suggests that our ability to experience romantic feelings has a genetic basis and is hardwired into our brains. In fact, recent neuroimaging studies conducted by anthropologist Helen Fisher and her colleagues pinpoint precisely the location of the "love circuits" in the human brain.[7]

Although Walt Disney movies feature animals that fall in love with each other or pets who love their owners, this anthropomorphic view of animal inner lives is far from reality. I believe that romantic love is unique to our species and probably evolved a few million years ago, after our ancestors split from the ancestors of chimpanzees and other apes. Like many of our other psychological, physiological, and physical traits, the human ability to love probably evolved by natural selection. That being said, traits that evolve by natural selection to serve a particular function can occur later on in contexts that have nothing to do with the original function.

Take, for example, sexual desire. No one would question that sexual desire and sexual attraction evolved to facilitate reproduction. Yet, in human societies, people are sexually attracted not only to individuals of the opposite sex, with whom they could potentially reproduce, but also to individuals of the same sex or to sexually immature individuals, and the reasons for this are complex. This doesn't necessarily mean, however, that there are different kinds of sexual attraction that evolved for different reasons. Similarly, romantic love is expressed in myriad different ways: children can fall in love with other children (I experienced my first crush at age seven), adults can fall in love with other adults of the same or the opposite sex, and both children and adults can be in love with their pets or even with inanimate objects such as toys or cars. That doesn't mean that there are many kinds of romantic love or that romantic love evolved for many different reasons. Regardless of its many expressions, romantic love is always the same phenomenon, and it is likely that it evolved to serve one particular function.

Consider this analogy. A hammer can be used to pound a nail, smash a window, or kill someone. That doesn't mean that there are different kinds of hammers or that hammers were invented for different reasons. Whoever invented the first hammer probably had only one function in mind—hammering nails—and the rest followed. The evolutionary function of love, I propose, is not to solve the commitment problem in cooperative relationships but simply to encourage men and women to form a bond that lasts as long as it takes to raise children together.

Pair-Bonding and Raising Children

In other primates—including apes, the ones most similar to us—sexual attraction, sexual intercourse, and reproduction work pretty much the same way as they do in humans, but there is no bonding between the mother and the father of offspring. Take orangutans, for example. When the time comes for reproduction, the male and the female meet for the first time a few minutes before they have sex and then never see each other again. When the baby is born several months later, the mother raises the baby on her own. Orangutan and other primate babies raised by single moms do just fine. Some get sick and die, of course, but that has less to do with the father's absence than with the baby's overall health. Also, the father's presence in these babies' lives wouldn't necessarily make them more successful as adults. Male orangutans, like many other primate males, impregnate females but don't help at all with the kids. There are no bad feelings between males and females about this, but there are no good feelings either. There is no attachment, or bonding, or anything remotely resembling romantic love between mates.

There are exceptions to this pattern. In a few primate species, the father's help is either necessary for infant survival or can make the difference between a bad life and a good one. In small South American monkeys called tamarins, females give birth to twins but are not strong enough to carry both babies around all the time. If the father didn't help with infant-carrying, the infants wouldn't make it past the first week. To keep the father around and ensure his help with the kids, males and females form pair-bonds and stick together for most of their lives. The tamarins and the orangutans exemplify a simple rule that is generally valid not only among primates but among all vertebrate animals: whenever one parent alone is enough to raise offspring successfully, single parenting is the rule and males and females don't form long-lasting pair-bonds.[8] (In most mammals the single parent is the mother; in fish species it tends to be the father.) However, when offspring need both parents to survive and become successful adults, males and females form pair-bonds and raise offspring together. Most species of birds are pair-bonded because chicks must be fed in the nest, and one parent alone could not do the job.

Psychologist Chris Fraley and two colleagues at the University of Illinois at Urbana-Champaign analyzed a large amount of scientific literature, first focusing on forty-four different taxonomic families of mammals and then on sixty-six species of primates, trying to reconstruct the evolutionary history of pair-bonding across time.[9] In their analysis, a species was defined as pair-bonded if mates spent a lot of time together and if there were signs of mate guarding, extensive physical contact and proximity, and separation distress. It turned out that mammals that were considered pair-bonded also provided bi-parental care; in addition, these animals had longer life spans and the period of offspring development was longer and slower. The analysis focusing only on primates found exactly the same results: both pair-bonding and bi-parental care are rare in primates, but when they occur, they go hand in hand. In these species, offspring develop slowly and are particularly vulnerable and needy. Fraley and his colleagues concluded that pair-bonded and bi-parental care co-evolved in mammals, including primates, and that in species in which fathers began contributing to child-rearing, pair-bonding soon followed.

Humans are no exception to the rule. If fathers did not cooperate with mothers in rearing children, children would lack important advantages they have when both parents help. Human fathers can provide help in important ways: they feed their children, protect them from danger, give them money, teach them all kinds of useful things, push them and support them, and get them out of trouble. There are two main reasons why humans are different from most other primates in the extent to which children need their fathers. One is that we have large brains, and the other is that we live in competitive and complex societies. Human brains are humongous by animal standards and take a long time to grow and mature. In fact, baby brains get so big during gestation that mothers must give birth before the brain has completed its growth; otherwise, the baby's head wouldn't fit through the birth canal. Brain development continues long after birth, and during this long, slow period of maturation human infants are vulnerable and need a lot of care. So a father's help can make a big difference. A father's help can also make a difference in ensuring that children become socially competent and successful adults. As I discussed in Chapter 2, because human societies are so competitive, young people need all the nepotistic support they can get to become successful adults. Even though single mothers can manage to raise their children and meet their basic needs, children who enjoy the additional support of their fathers have double the nepotistic connections to rely on (especially if they end up in the military or in academia).

If male-female pair-bonds in animals and humans are about raising children together and not only about founding film production companies, as Brad and Jennifer did, and if love exists to cement this bond, then one might expect that love would fade and the bond dissolve if children are not conceived (building a nursery in the new house is not enough—you actually have to put a baby in it) or when the parents' job is done and they want to move on with their lives.

Breaking the Bond

In the 1955 film *The Seven Year Itch*, a married man struggles with the temptation to leave his wife and small child to run off with the young woman next door, played by Marilyn Monroe. (This movie contains the famous scene in which Monroe stands on a subway grate and her dress is blown above her knees by a passing train.) The title of the film refers to the time in a marriage when—according to the U.S. Census Bureau—a divorce is most likely to happen. One possible explanation for this timing is that after seven years of marriage many couples have successfully raised one or two children through the risky infancy years and can catch their breath and realize they don't want to be around each other anymore, or else they haven't had children at all and decide it's time to look for another potential mate. Brad Pitt and Jennifer Aniston divorced after having spent seven childless years together, so the timing of Brad's itch is consistent with this theory. Or perhaps when Marilyn Monroe or Angelina Jolie enter the picture all bets are off and, kids or no kids, married men (including presidents of the United States) start itching and getting ideas in their heads. I can cite some research in support of the first theory; support for the second theory can be found in Marilyn's and Angelina's biographies.

To understand the first theory, it is important to know that the risk of infant mortality in humans and other primates is very high in early infancy and decreases steadily as infants grow older. Thus, if a couple splits up shortly after a baby is born, the baby's survival or well-being could be at risk. Anthropological study of indigenous cultures supports the notion that in the first few years of life human children need all the help they can get. Breast-feeding is generally known to reduce the baby's risk of catching diseases because the mother transfers her immune defenses against germs to the baby through her milk. In contemporary hunter-gatherer populations such as the !Kung in South Africa, mothers breast-feed their babies for approximately four years. Emory University anthropologist Melvin Konner, who studied the !Kung in the 1970s, has suggested that four years was probably the average interbirth interval for humans during much of our evolutionary history.[10] Among the hunter-gatherers, a baby is fully weaned at four years old: at this age, the transition from breast milk to solid food is complete. The risk of infant mortality then drops even further. At this point, the mother typically has another child. But not all couples proceed to have a second child.

In the late 1980s, anthropologist Helen Fisher, author of the 2004 book *Why We Love: The Nature and Chemistry of Romantic Love*, gathered divorce data from fifty-eight different human societies around the globe from the records of the *Demographic Yearbooks* of the United Nations.[11] She discovered that when married couples divorce, they tend to do it around the fourth year of their marriage, typically after having had a single child. One interpretation of this discovery is that many human couples stay together at least the minimum amount of time necessary to successfully raise one child together. Fisher took this idea a step further, however, and speculated that humans might have a predisposition to be serial monogamists. Although "serial monogamist" sounds a lot like "serial killer," it simply means that people are socially bonded to one partner at a time, but don't stick to the same partner their whole life; they go from one partner to another, in succession. According to Fisher, humans are likely to switch partners every four years, after having a child. In reality, there is no strong evidence that humans are serial monogamists. Although the number of divorces keeps increasing in many societies around the world, there are many couples who remain married forever, raise children together, and aren't convinced that their parenting job is done even when it's pretty clear that their children can make it on their own. My own parents have been married for over fifty years, and my mother still insists on buying me socks and underwear. Clearly, she doesn't think her job is done.

When divorce does occur, the interval between marriage and divorce varies widely across societies and historical periods. A 2010 study of marriages and divorces in Italy conducted by ISTAT, a research group that conducts statistical analyses of demographic data, found that Italian marriages last, on average, fifteen years.[12] The average age at which men get divorced is forty-five, and for women it's forty-one. Italians now marry relatively late, in their mid and late thirties, so divorce tends to occur when couples have either been married for several years and produced no children or when they have had one or two children and the youngest is four to five years old. Again, this is consistent with the theory that humans form pair-bonds in order to have and raise children.

Common sense suggests that when married couples divorce, there either wasn't much love to begin with or love was strong at the beginning but then gradually weakened. The latter possibility is consistent with another theory of love, developed by a friend of mine and related to me after a few drinks, according to which love is just a stage in a relationship—an early stage. My friend's theory is actually quite compatible with the theory about pair-bonding and child-rearing. Unlike Robert Frank's commitment theory, according to which love between two partners should grow stronger over time, as the risk of defection increases, the "early stage" theory and the "pair-bonding and child-rearing" theory both maintain that love must be strong at the beginning of a relationship and last at least a few years, but doesn't necessarily need to remain so forever.

But if humans form pair-bonds to jointly raise their children the way birds and other animals do, why is it that romantic love occurs only in our species? Cooperation between mothers and fathers poses the commitment problem, and Frank suggests that romantic love evolved to solve this problem. Is cooperation in rearing human offspring in some way different from other forms of animal or human cooperation such that it presents unique problems requiring unique solutions? I don't think so. As a cooperative partnership, the human pair-bond is not qualitatively different from other long-term cooperative relationships. In many species of animals, pairs of individuals cooperate in finding food, keeping each other clean, or supporting each other in fights against other individuals. Humans cooperate with other humans in a million different contexts. All of these cooperative relationships pose the commitment problem, yet the problem is not solved with romantic love. As we'll see in the next chapter, commitment problems are not addressed with irrational feelings but with repeated testing of the bond.

Love is an emotion, so to understand its evolutionary function we have to place it in the broader context of the evolutionary function of emotions. Although emotions occur in the context of social relationships, they also occur in many other nonsocial contexts.

Love the Energizer

When one person joins another
It creates a chain reaction, which leads to fission
The release of energy culminating in a vast explosion
Whose result is total devastation
The wiping away of cities
The crumbling of continents
The destruction of mountain ranges

The melting of glaciers
The boiling away of oceans
The cracking of a planet
The collapse of a solar system
A galaxy folded in upon itself
The universe, gone
Just because two people get involved.

—Joe Frank, "Love Prisoner"
monologue from his radio show *The Other Side*

One major function of emotions is to energize motivation. If we experience a strong positive or negative emotion, we become motivated to do something that's beneficial or to avoid something that's harmful. Pain exists to make sure organisms do everything they can to avoid things that can damage their bodies. If you are crazy enough to stick your finger in the flame on the kitchen stove, pain is there to protect your body and make it difficult for you to hurt yourself, no matter how crazy you are. Sexual desire and orgasm exist to make sure organisms are highly motivated to engage in sexual intercourse and produce children, regardless of their opinions on the subject. Sexual urges are so powerful that it is difficult even for priests who take a vow of celibacy to completely suppress their sexuality. Their emotions work against their conscious decisions, and the result is that some of them end up in the news for engaging in inappropriate sexual behavior. People make arbitrary decisions about all kinds of things, but matters of survival and reproduction are too important to be entirely dependent on people's conscious decisions. Emotions evolved to encourage us to do what's good for us regardless of what we think about it. Romantic love evolved, I argue, to motivate men and women to form pair-bonds.

But why the need for this extra emotional energizer? I think the answer has to do with our primate evolutionary past. In birds pair-bonding is an ancient adaptation. Birds have probably been pair-bonded organisms for many millions of years. This means that natural selection has had plenty of time to sculpt birds' brains and provide the necessary wiring to support the psychological and behavioral adaptations for pair-bonding. In comparison to birds, human pair-bonding is an evolutionary novelty. It arose very recently—a few million years is equivalent to the day before yesterday on the evolutionary time scale—and very quickly in response to the rapid changes in brain size and patterns of child development that made bi-parental care necessary or advantageous. To complicate things, humans probably evolved from a chimpanzee-like ape species whose members were sexually promiscuous, whose males did not contribute anything at all to child-rearing, and in which there were high levels of conflict between the sexes (such as male aggression toward females and sexual coercion). The brains of our ape ancestors had probably been shaped by sexual selection for millions and millions of years to support mating and reproductive strategies that did not involve pair-bonding. As psychologist Paul Eastwick argued recently, when the circumstances became favorable for the evolution of pair-bonding in the human lineage, natural selection had to quickly modify human brains in ways that would counteract other features that had been honed through eons of sexual selection.[13] It wasn't an easy evolutionary step for the male brain of a sexually promiscuous, aggressive, and misogynistic chimpanzee-like ape to become the socially monogamous, female-loving, and paternal brain of a human being. The need for this rapid transformation presented a special evolutionary problem, which required a special solution. This special solution was romantic love and adult attachment.

But how did natural selection find this special solution? How did it come up with romantic love?

The History of Love

Being often at the airport, I cannot help but notice how people behave when they say good-bye to their loved ones. I have seen many a husband hold his wife's hand while she is getting her boarding pass at check-in and carry her baggage until she gets past the security line; finally, before they separate, they smile, hug, and kiss each other. I have also seen tears in both partners' eyes and other painful expressions when they finally have to let go of each other's arms. I am clearly not the first or the only behavioral scientist who has observed these scenes and wondered about the nature of love. Psychologists Chris Fraley and Phil Shaver conducted a study of this phenomenon, which was published in 1998 in the *Journal of Personality and Social Psychology* with the title "Airport Separations: A Naturalistic Study of Adult Attachment Dynamics in Separating Couples."[14] They conducted this study because they thought that the way romantic partners behave during airport separations might shed light on the nature and origins of human love.

Fraley and Shaver had a team of four observers who took written notes on the behavior of separating couples at a small metropolitan airport. For comparison purposes, the four observers also recorded the behavior of couples who were flying together. These are some of the behaviors they observed in the separating couples:

> They held hands.
> They hugged and held each other for about five minutes.
> He kissed her several times when she tried to leave.
> They gave each other a long and intimate kiss.
> He massaged her inner thigh.
> They both cried and wiped the other's tears away.
> She, in a comforting manner, stroked his face.
> He walked away quickly; she walked away crying.
> She whispered, "I love you," to him as she boarded.
> They continued to hold hands even though they
>   were walking away from each other.
> At departure, she was the last to board the plane.
> After he left, she went back to the window and
>   watched the plane leave.
> She was still at the window twenty minutes after the plane left.
> In one case, a man who had already boarded the plane came
> rushing out for one last kiss, unleashing the wrath of the flight
> attendants, who urged him to return to his seat immediately.

In addition to observing and recording the behavior of these separating couples, the researchers also asked them to fill out questionnaires about their personality, the length of their relationship, and the degree of subjective distress they experienced during separation from their partner. The main goal of this study was to show that separating couples engage in behaviors functionally similar to those observed in children who are being separated from their parents. In both cases, there are behaviors that function to seek or maintain contact or proximity, expressions of sadness or distress, caregiving and comforting behaviors, and sometimes aloofness and rejection of affection. The general theory inspiring this study, which also guides much of contemporary research on romantic love, is that love and adult attachment have their evolutionary origins in the emotional and social bond between a child and its caregiver. The bases for this theory were laid out about fifty years ago.

In the early 1960s, British psychiatrist John Bowlby developed a theory to explain why young children are strongly bonded to their primary caregiver, usually their mother. According to Bowlby's "attachment" theory, young children have a biological predisposition to become emotionally attached to a caregiver, which manifests itself as behaviors aimed at maintaining proximity or stimulating interaction, such as crying, smiling, following, and clinging. Human infants are totally dependent on their caregivers for protection from the environment. Bowlby argued that the attachment system probably evolved by natural selection as a set of psychological and behavioral adaptations that promote infant survival by enhancing infant proximity and interaction with a caregiver. Bowlby got his ideas about attachment from observing children's reactions to separations from their parents before being checked into a hospital, as well as from reading descriptions of mother-infant interactions in monkeys. It turns out, in fact, that the infant attachment system is not unique to the human species but exists in many species of monkeys and apes that are closely related to humans. It's at least 35 million years old—a lot older than human pair-bonding.[15]

Bowlby argued that the infant attachment system has a set goal—the maintenance of contact with or proximity to the mother—and specific activating and terminating conditions. The attachment system is activated when the infant is separated from the mother, and it's terminated when contact or proximity is achieved. The system thus works like a thermostat that measures the current temperature and after comparing it with a preset standard, makes adjustments. The infant attachment system has three defining features: young children show anxiety and fear of strangers when separated from their mother (they cry, cling, and become anxious or sad); they run back to their mothers when they are scared (using her as a "safe haven"); and when they are calm and confident they use their mother as a "secure base" for exploration, walking away from her to explore and play but often checking back with her to make sure everything is okay.

Although infants cry from the moment they are born, the main defining features of the attachment system, such as separation anxiety and fear of strangers, first appear when children are between six and nine months old. The attachment system continues to operate through the rest of childhood and even into adolescence, although the behaviors through which attachment is expressed may change. The basic features of the attachment system can be seen in all children, although not all children react the same way to a short separation from and subsequent reunion with their mother; some children are more comfortable being separated and are loving and affectionate at the reunion, whereas others get very anxious during separation and express anger and rejection toward the mother at the reunion. It turns out that both the basic features of infant attachment and the differences among children can be observed not only in all human societies and cultures around the globe, but also in the species of monkeys and apes that are most closely related to us: rhesus monkeys, baboons, and, of course, chimpanzees and the other great apes.

Freud was among the first to write about the striking similarities in the physical intimacy that typifies lovers and mother-infant pairs, including the way romantic partners use baby talk with each other.

Fraley and Shaver's airport study and many others conducted since the late 1980s have shown that the features that characterize the infant attachment system—proximity maintenance, separation distress, safe haven, and secure base—can also be observed in romantic relationships between adults.[16] Moreover, children who were securely attached to their mothers turn into adults who are comfortable and laid back in romantic relationships, whereas children who were insecure, anxious, and ambivalent/angry toward their mothers (or fathers), well, they behave the same way toward their romantic partners in adulthood. Finally, there are also many similarities in the temporal development and transformation of attachment relationships in mother-child pairs and romantic partners. Romantic attachments are usually more mutual and symmetrical than child-mother attachments. In addition, in romantic relationships partners often alternate in the roles of the immature individual who needs attention, comfort, and reassurance and the caregiver who provides all these things.

Given all of these theories and observations of the last fifty years, I suggest that the evolutionary history of human romantic love may have progressed along the following lines. As human brains grew and infants became needier and more vulnerable for a longer period of development, such that the father's involvement and bi-parental care became necessary, natural selection had to come up with a way to motivate men and women to stay together for as long as it took to raise a child successfully. Now, natural selection never invents anything from scratch but rather modifies and rearranges preexisting structures. The psychological and emotional adaptations for the infant attachment system already existed in the brains of our ape ancestors and had worked pretty well to keep infants and mothers together. Natural selection tweaked this system, making it operational through adulthood, so that it could be used to bond mates to each other. Some of the neural circuits and the neurochemical substances that had been used to bond mothers and children, such as those involving oxytocin and endogenous opioids (which are also involved in regulating the body's responses to stress and physical pain), also became involved in mediating bonding between adults.

To accomplish the goal of fostering long-lasting emotional and social bonds between adult males and females, natural selection tinkered not only with the brains of our ape ancestors but with their bodies as well. The bodies of our ape ancestors were probably similar to the bodies of modern chimpanzees: well adapted for intense sexual competition and sexual conflict, but not for pair-bonding. For example, males were larger and stronger than females, had larger and sharper canine teeth, and had relatively small penises but huge testicles that produced large quantities of testosterone and sperm. Females, for their part, advertised their fertility period during their menstrual cycle through large sexual swellings to incite sexual competition among males. To foster pair-bonding and cooperative relationships between the sexes, natural selection reduced the differences in body size, strength, and weaponry between males and females. Then it eliminated obvious signs of ovulation in women and increased their receptivity throughout their menstrual cycle. This provided the opportunity for paired men and women to have sex all the time, thus reinforcing their union and increasing the man's confidence that when a child was born it was really his, which in turn increased his willingness to provide paternal care. At the same time, natural selection reduced paired men's desire for sexual variety and promiscuity by reducing their testis size and lowering their testosterone levels. Human males have relatively small testicles for their body size and produce small amounts of sperm and testosterone compared to male chimpanzees. I once saw a slide of a researcher holding a chimpanzee brain in one hand and a testicle in the other; they were approximately the same size, and not because chimpanzee brains are small.

Another physiological adaptation for pair-bonding in human males is the dramatic reduction in their production of testosterone when they find themselves in committed relationships or are married with children.[17] Lower testosterone in romantically committed men curbs their desire for other women and allows them to concentrate on their wives and children. This has been shown by many studies, including one that my colleagues and I conducted at the University of Chicago involving over five hundred MBA students. Finally, various researchers, including psychologists Cindy Hazan and Debra Zeifman, have suggested that the exceptional length of the erect human penis—human males have by far the longest penis in relation to their body size of all the primates—is also an adaptation for pair-bonding.[18] The long penis makes possible a wide variety of copulatory positions, including more intimate face-to-face, mutually ventral positions, which promotes social bonding during sexual intercourse. Ventro-ventral sexual intercourse is rare in primates but common in another species closely related to us, the pygmy chimpanzee, or bonobo; like humans, bonobos use sex for social bonding purposes. The long human penis may also increase the probability of female orgasm, which heightens the female's readiness for engaging in sexual activity, thereby strengthening the bond with her mate.

The multiple physical, physiological, and psychological adaptations that have arisen through natural selection to induce human males and females to form pair-bonds and cooperate in rearing offspring generally work very well. The most amazing psychological adaptation for pair-bonding—romantic love—creates in the human mind a longing for the desired partner and a psychological dependence not dissimilar from that existing between a young child and her mother. Successful bonds involve a profound psychological and physiological interdependence between partners such that the absence or loss of one partner can be literally life-threatening for the other. Conversely, solid and stable romantic relationships can have many positive effects on the health and longevity of both partners and their children. Although economics can help us understand some aspects of pair-bond formation and some of the issues involved in cooperation between partners, evolutionary biology tells us that human romantic pair-bonds are far more than business partnerships based on the principles of cooperation and reciprocal altruism.

So back to the original question: Why did Jennifer Aniston and Brad Pitt first fall in love with each other only to break up? Can the theories from economics and evolutionary biology discussed in this chapter explain why some Hollywood celebrities (as well as us common mortals) form pair-bonds and then split up?

Well, read the following quotes from the August 30, 2010, issue of US Weekly and judge for yourself:

Actor Neil Patrick Harris [the famous TV child doctor Doogie Howser] and his partner David Burtka are expecting twins via a surrogate mother. A source close to the couple reveals: "They bonded about wanting kids and having a family from the very beginning, which is why they fell in love. It's a dream come true."

Actress Halle Berry, speaking about former partner Gabriel Aubry, with whom she has daughter Nahla, 2: "You realize you are not meant to go the distance with everybody. We were meant to bring this amazing little person into the world . . . and we are a family until we are not here anymore."

Chapter 7

Testing the Bond

The Baboon Solution to the Commitment Problem

Male baboons form cooperative relationships with each other that involve aggressive alliances and mutual support in fights. Like the business and romantic partners in the previous chapter, male baboons, too, face the commitment problem: one of the two partners might cheat and double-cross the other, or simply end their relationship, leaving the other in the dust. So they've come up with an unusual way to deal with the commitment problem: they fondle each other's testicles.

Other primates are also known to engage in similar practices. In ancient Rome, two men taking an oath of allegiance held each other's testicles; men held their own testicles as a sign of truthfulness while bearing witness in a public forum (hence the word "testify"[1]). The behavior of male baboons and ancient Romans can be explained by the Handicap Principle (HP), a biological theory according to which the most effective way to obtain reliable information about a partner's commitment in a relationship—whether a friendship, a romantic relationship, or a business partnership—is to impose a cost on the partner and assess the partner's willingness to pay it. But before I say anything more about this, let's go back to the strange habits of male baboons.

Adult male baboons are aggressive and dangerous animals. Despite their relatively small size—approximately that of a German shepherd—male baboons are strong and fast, with large, sharp canine

171

teeth that make their bites potentially deadly. They are also fearless and won't hesitate to attack animals much larger than themselves, such as lions and chimpanzees. But what makes male baboons so dangerous is that when they attack they are rarely alone; they do so in pairs or small gangs of three or four males. Although male baboons cooperatively hunt smaller animals when they feel like eating meat for dinner and also cooperatively attack large predators that prey on baboons, both of these reasons for cooperating are relatively rare. When male baboons form aggressive alliances, most of the time they do it to fight other male baboons.

Baboon society is a competitive environment. Adult males compete with each other to mate with females and to acquire and maintain high-ranking positions in the dominance hierarchy. Baboons are highly promiscuous monkeys—males like to mate with many females, and often. Male baboons have large testicles for their body size and can ejaculate ten to fifteen times a day, or more. High-ranking males, however, monopolize fertile females and won't let lower-ranking males come near them. The 1977 study by Craig Packer that I mentioned in Chapter 2 showed that male baboons have figured out a way to circumvent the monopolization of fertile females by high-ranking males.[2] Two males will work in tandem: while one of them picks a fight with the high-ranking male who is guarding the female, the other one takes advantage of the diversion to mate with her. A few days later they play the same game but with reversed roles: the guy who got lucky with the female will return the favor to his buddy and pick a fight to create the diversion. Since fighting to help another individual mate is an altruistic act, Packer's study has become a classic example of reciprocal altruism in primates. Adult male baboons also form coalitions when they fight for power, just like the male macaques discussed in Chapter 4, but even more so. To rise in rank, two or three adult males may gang up against a higher-ranking male, defeat him in battle, and make him lose his status or kick him out of the group altogether.

Whether male-male coalitions are formed to gain sex or power, they are not formed instantly or randomly. Just as the formation of human political alliances requires a lot of networking, baboons will take a long time to get to know each other, form a social bond, and develop a certain degree of trust before they form an alliance. The social bonding between coalition partners entails hanging out together and exchanging a lot of grooming. If adequately maintained, cooperative relationships between two males can last a long time and result in many coalitions.

But just as there is competition in mating and dominance, there is also competition for good political allies, and high-ranking males are sought-after coalition partners. This competition makes cooperative relationships between males unstable, capable of ending at any time, and ripe for betrayal: even the male baboon that you thought was your best friend could, in the middle of a fight, decide to switch allegiances and side with your worst enemy. Male baboons cannot sign a written agreement to cooperate, much less enforce the contract with the threat of financial sanctions. Nor can baboons talk, so a male baboon must figure out in some other way whether his friendship with another baboon is strong or weak and whether the other party can be trusted as a coalition partner.

Now imagine this scene. Forty to fifty baboons are sitting around and enjoying the breeze on a sunny summer afternoon. Some monkeys are looking for food on the ground, while others are grooming or sleeping. Clint Eastwood and Eddie Murphy are two adult males who occupy high-ranking positions in the group and are quite successful with the ladies. Each has a small harem of three or four females, with whom they mate all the time and who are loyal to them. Other males in the group are lucky to have one female who pays attention to them, and some old guys have been out of the mating business for a while. (In this particular species of West African baboons—the Guinea baboon—males keep small harems of females, a bit like the hamadryas baboons that live in East Africa.) Clint and Eddie are sitting twenty yards apart, each minding his own business. Out of the blue, each glances at the other over his shoulder and makes eye-to-eye contact. In that split second, their baboon minds decide that it's time for some bond-testing.

Clint and Eddie run toward each other, meeting approximately halfway. As they run, Clint makes a funny facial expression, his eyes half-closed and his ears flattened back against his skull, while Eddie smacks his lips very rapidly. When they meet, each baboon lifts his leg—the way a male dog does to pee—and the other briefly holds his testicles in his hand. Without even looking at each other, they run back to where they were before, sit down, and continue with their business as if nothing had happened. The whole interaction lasts mere seconds and happens spontaneously: nothing has led to it, and nothing follows. The other baboons in the group hardly notice—they've seen this scene a million times before.

My former PhD student Jessica Whitham saw hundreds of these interactions (and captured some good ones on video) when she observed the group of Guinea baboons at the Brookfield Zoo in Chicago to collect data for her master's thesis.[3] (Incidentally, she is the one responsible for naming the baboons after Hollywood celebrities.) Although homosexual behavior is not uncommon in monkeys and apes—both males and females mount individuals of the same sex under various circumstances—the fondling of genitalia performed by male baboons has nothing to do with sex. It's a social ritual that primatologists call a "greeting." In her thesis, Jessica showed that pairs of adult males who have stable cooperative relationships—characterized by frequent grooming and alliances—often exchange greetings, while males with poor or unstable relationships either don't exchange greetings at all or initiate greetings but are unable to complete them.

In an incomplete or aborted greeting, one male winks at another one and starts walking toward him, but when the other turns the other way and does nothing, the first male stops and goes back to where he was. Sometimes the two males approach each other, but a second before grabbing each other's testicles they change their minds, then quickly turn around and retrace their steps. It's as if one or both of them thought at the very last second that this intimate exchange wasn't such a good idea after all, or freaked out about how the other one might react. Another researcher who has studied greetings between male baboons, University of Michigan primatologist Barbara

Smuts, discovered that of over six hundred greetings she observed, about half were incomplete because one male, or both, pulled away before completing the exchange.[4] The interruption of the sequence could occur at any time, ranging from right at the beginning, when one baboon initially glanced at the other, to almost at the end, when his hand was beginning to touch the other's testicles. Observing the interaction on slow-motion videotapes, Smuts discovered that during the course of a greeting two baboons monitor each other and respond to the subtlest glances and shifts in movement with split-second timing. Any sign of hesitation in the other partner can be a reason for terminating the greeting before it's complete.

When two baboons fondle each other's genitalia, they take a huge risk. Each baboon could quickly and easily terminate the other's reproductive career for good by ripping his testicles off. Thus, letting another baboon fondle one's testicles shows a great deal of trust in the other's good intentions. On the other hand, by getting so close to another male and attempting to touch his testicles, a male exposes himself to a high risk of aggression. A single bite inflicted with a male baboon's sharp canines could scar someone for life. Again, initiating a greeting requires a great deal of trust that the other individual will not respond aggressively to this potentially dangerous violation of privacy. Greetings work only if both males are equally committed to being friendly. When friendly feelings are not fully reciprocated— for example, when a low-ranking male tries to befriend the alpha male but the alpha is uninterested—greetings are attempted but aborted.

Occasionally greetings end in a fight. Smuts reported that about 7 percent of the greetings she observed ended in threats, chases, or physical aggression. Pairs of very close friends show the most intense greetings—those in which both the penis and the testicles are fondled for a few seconds—and their greetings are never incomplete. When relationships are good, it doesn't really matter if one partner ranks higher than the other. Generally, low-ranking males initiate greetings about as often as high-ranking males do. A low-ranking male may use a greeting to gather information on the dominant's willingness to tolerate and support him. A high-ranking male may use a greeting to communicate his willingness to accept a low-ranking male's presence in his group and his availability as a potential ally. The two partners in the cooperative relationship frequently assess each other's commitment to it—testing the strength of their bond—by imposing on each other. What makes the greeting work as a bond-testing mechanism is the risk associated with it—and therefore the potential cost that each individual is willing to pay to maintain the relationship. By taking the risk and tolerating the imposition, male baboons demonstrate how much they value their relationship. Let's now explore in more detail why this is the case.

The Logic of Bond-Testing and the Handicap Principle

The importance of bond-testing was highlighted in "The Testing of a Bond," a paper published in 1977 by an Israeli evolutionary biologist, Amotz Zahavi, who presented some novel and controversial ideas about cooperation and communication.[5] As an evolutionary biologist, Zahavi reasoned that cooperative relationships between individuals who are not genetically related to one another are intrinsically less stable than those involving family members. Two unrelated individuals, as we know, may form a cooperative relationship because they have shared interests and want to jointly pursue goals that would be difficult or impossible to accomplish without a partner, such as having children. Zahavi recognized, however, that the circumstances that hold a cooperative relationship together can change quickly and unpredictably. Therefore, he emphasized, it is important to frequently test the strength of the bond and assess the partner's commitment in order to decide whether to continue investing or bail out.

Take human romantic relationships. The most direct way for romantic partners to assess their mutual commitment to the relationship is simply to ask each other all the time, "Do you love me?" "Are you sure you love me?" or "Are you sure you want to be with me forever?" Couples in love do this all the time, as they should, but unfortunately this is not the most reliable of methods for assessing commitment (and for animals, it's not an option at all). People can be insincere, or even clueless, about their feelings and future behavior. Zahavi's original idea was that the most reliable way to assess how much a relationship is worth is to assess its market value, that is, how much someone is willing to pay for it. Your boss at work can tell you that you are a valuable employee and praise your work constantly, but the best indicator of how valuable you are to your boss is the salary he or she is willing to pay you. Words are cheap, but money isn't.

When money is not involved, the price one is willing to pay for a commodity can be counted in a different currency. For animals, the most meaningful currency is fitness: the ability to survive and reproduce. This is the currency that natural selection does business with. To assess the value to an animal of a commodity (such as food, a mate, or a relationship with an alliance partner), one has to measure the extent to which the animal is willing to risk its survival or future reproduction to obtain or keep that commodity. According to Zahavi, "The only way to obtain reliable information about another's commitment is to impose on that other—to behave in ways that are detrimental to him or her." Zahavi believes that the information about commitment obtained with this bond-testing mechanism is reliable because only partners who are truly committed will accept the imposition. By gradually escalating the imposition and determining the point at which the partner will terminate the interaction, one can obtain precise information about the other's commitment to the relationship and current willingness to invest in it. A male baboon who wants to know exactly how much his alliance partner values their relationship can continue holding his partner's testicle until he gets smacked in the head or bitten. The probability of getting a negative reaction increases exponentially with time, so being able to prolong the ritual for even one second is a significant accomplishment that bespeaks the strength of the commitment.

The idea that the strength of a social bond can be tested by imposing a cost on the partner is one application of the more general theory known as the Handicap Principle, which Zahavi first presented in 1975 and expanded and refined in subsequent years.[6] The HP was developed to explain the existence of honest or unreliable information in animal communication, but it could be applied to many other phenomena as well, including, as we'll see later, many aspects of human behavior. The main idea of the HP is that a cheap signal is easy to fake and anyone can do it, while expensive signals require resources that only superior individuals possess. Therefore, expensive signals convey honest information about the quality of the individuals who are able to use them.

Zahavi purportedly had the idea for the HP while attempting to explain to a student why peacocks have giant and elaborate tails and why peahens prefer to mate with males with the longest, heaviest, and most cumbersome tails: if large tails were cheap to produce, then all males would have them, regardless of age, size, strength, or health. But a large tail ain't cheap and may actually be detrimental to the male's survival: it's expensive to produce and can hamper the peacock's mobility, potentially reducing his ability to escape a predator. Strong, healthy males can afford huge tails, while weak and sick males can't. In Zahavi's terms, the peacock's tail is a handicap. So why would a male want to take on such a handicap? Perhaps simply to show the female that he can. The peacock is self-handicapping to show the peahen how good he is. The bigger the handicap, the better the male.

Thus, according to the HP, individuals signal their quality by displaying traits that are detrimental to themselves. Signals with handicaps are inherently honest owing to their cost; females trust males who display such signals and find them attractive as mating partners. Of course, peacocks who handicap themselves with a large and expensive tail don't do it by conscious choice; sexual selection, not conscious thinking, is responsible for the evolution of large tails. The males with the genes for such tails are more attractive to females, reproduce more than other males, and therefore leave more copies of their genes in the population.

The peacock's tail is an example of a physical handicap, but handicaps can be behavioral as well. Behaviorally, a handicap can entail taking risks that may reduce one's probability of survival. One classic example of a behavioral handicap is the stotting behavior displayed by African gazelles when confronted by a predator, such as a lion. When gazelles spot a lion, some of them start jumping up and down in front of him instead of fleeing as fast as they can. Why do these animals waste time and energy on behavior that could jeopardize their life? Zahavi's answer is that stotting is a handicap and that some gazelles take it on to communicate how strong and fast they are. They are telling the lion that they would be difficult to catch and therefore a waste of time. The lion would be better off chasing the easier prey—the other gazelles that are trying to get away as quickly as they can. The assumption here is that lions sometimes do chase the stotting gazelles, or at least did so in the past when stotting behavior first evolved. The gazelles who cheated and stotted even though they were not fit enough to do it paid with their lives, and their genes were not passed on to the next generation. Over evolutionary time, the genes for cheating were wiped out by natural selection while the genes for honest stotting spread in the population. (Note: This example is an oversimplification to illustrate how natural selection works. In fact, specific single genes for "honesty" or "deception" don't exist, and behavioral traits are often controlled by multiple genes interacting in complex manners.)

Another, less well-known potential example of behavioral handicapping in the animal world involves the bizarre behavior of adult male baboons holding an infant in the middle of a fight.[7] Imagine this scene. Two male baboons are about to have a fight: they threaten each other by staring each other down and ominously displaying their canines. All of a sudden, one of the males grabs a nearby infant—sometimes forcibly removing it from its mother's arms—then continues to threaten his opponent, holding the infant in his arms. Why does he do this? Is he telling the other guy that he is a father and begging for mercy? Is he threatening to harm the other's son? Is he trying to use the infant as a shield? The HP suggests another explanation: fighting while holding an infant in one's arms is clearly a handicap, so by grabbing and holding an infant, a male wants to tell his opponent that he is so strong that he can beat him in combat even with an infant in his arms. It's the primate equivalent of: "I can beat you up even with my hands tied behind my back."

For all three examples—the peacock's tail, the gazelles' stotting behavior, and the baboons' infant holding—there are other possible explanations, though it has proven difficult for researchers to establish which of them are correct. But the explanation provided by the HP is interesting and makes sense; until it is tested and unequivocally discarded with appropriate data, it is as viable as any of the others.

According to Zahavi, the HP explains not only costly and extravagant morphological traits and risky and bizarre behaviors in animals but also many general social phenomena, including altruism and nepotism. Zahavi argues that when individuals behave altruistically they are not really trying to help others. Instead, they handicap themselves to advertise their own quality and resources. Self-handicapping for the purpose of advertisement works even better when money or resources are not donated to others but simply wasted, as in the human phenomenon of conspicuous consumption. Rich people like to waste their money on luxurious yachts or fancy cars, not because they need them but simply to show off. University of New Mexico evolutionary psychologist Geoffrey Miller has written a book, called *Spent: Sex, Evolution, and Consumer Behavior*, in which he argues that many forms of wasteful spending in contemporary human societies are manifestations of the Handicap Principle.[8] Wealthy men throw their money away to make themselves attractive to potential mates, and both men and women waste money on luxury goods to gain social status with their peers. Again, there are other explanations for acts of altruism such as wealthy people's donations to charities, some of which I discussed in Chapter 5. And many economists would probably object to the notion that the HP and its role in mate attraction fully explain consumer behavior in capitalistic societies. However, I suspect that economists—at least those at my parent institution—would be generally sympathetic to the reasoning at the basis of this principle. The HP paints a view of the world in which living organisms are fundamentally rational and selfish creatures and their social interactions are governed by the rules of the market and the advertising business. The HP explains the natural world using the principles of capitalism.[9]

Although there are many similarities in the cost-benefit analyses used by economists and evolutionary biologists, as well as an increasing awareness that these disciplines show a great deal of convergence in their general principles, back in the 1970s, when the HP was first proposed, this kind of reasoning was radical in evolutionary biology. By proposing economic cost-benefit analyses of animal signals, Amotz Zahavi was way ahead of his time. And as it may be easily imagined, his ideas were met with a lot of skepticism and resistance before they were accepted, or even taken seriously, by other evolutionary biologists. In fact, he continues to fight this battle after thirty-five years now. Whether he has won or lost the battle depends on who you talk to, but one thing is certain: he has managed to get a lot of people's attention in the process. His personal story and the vicissitudes of the HP nicely illustrate, in many ways, how the process of scientific research works and some of the dynamics between researchers I discussed in Chapter 5.

At the time of this writing, Amotz Zahavi is eighty-two years old and an emeritus professor of zoology at the University of Tel Aviv in Israel. Despite being formally retired from academia, he is frequently invited to speak at conferences or major universities around the world. He speaks English with a strong foreign accent and comes across as a bit grumpy and old-fashioned: for example, he often distributes handwritten notes to his audience instead of using a computer and PowerPoint for his presentations. He also presents his ideas very forcefully and quickly gets defensive when people ask him questions. In his seminars and articles, Zahavi never fails to mention that when he first presented the HP it was unanimously rejected. According to him, the prominent British evolutionary biologist John Maynard Smith agreed to publish his 1975 article in the *Journal of Theoretical Biology* only to write and publish, less than one year later, a paper of his own to prove that the HP was all wrong. Many other researchers published articles and commentaries that criticized and rejected the HP. According to Zahavi, the main reason for skepticism was that

Figure 7.1. Dr. Amotz Zahavi. Photo from Wikipedia.

he had initially formulated and expressed the HP in qualitative terms, intuitively, without backing it up with appropriate mathematical models that would show how genetically determined morphological and behavioral traits could evolve and be maintained in populations through the HP. Some evolutionary biologists liked the idea of the HP, but didn't think it actually worked in the real world. Then, in 1990, a well-regarded evolutionary biologist and mathematician at Oxford University, Alan Grafen, showed mathematically that the Handicap Principle was viable and that this mechanism was consistent with the dynamics of Darwinian evolution. Since then, according to Zahavi, the HP has become widely accepted.[10]

In reality, the HP is still controversial and being debated; even the supporting articles written by Alan Grafen have been criticized. Many evolutionary biologists believe that some aspects of the HP are helpful in understanding how communication works and why signals are the way they are. But many are also skeptical about the general validity and applicability of this theory to a wide range of biological phenomena and do not believe, as Zahavi does, that the HP should replace other well-established theories in evolutionary biology, such as reciprocal altruism or kin selection theory.

As often happens in science, when researchers develop a new theory or a new research approach and think that it can effectively explain some previously unexplained phenomena, they get excited about it and try to apply it to everything else. They announce that the new paradigm will replace old ones and eventually take over entire fields of research or disciplines. This has happened with other paradigms in the behavioral sciences, such as behaviorism and sociobiology. Psychologist B. F. Skinner (1904–1990) was a major supporter of behaviorism—the paradigm that emphasizes the role of learning in behavioral phenomena—and came to believe that the principles of conditioning should be applied to all aspects of human life, from raising children to government. When he tried to argue, in a book called *Verbal Behavior* (1957), that human languages are learned just like any other aspect of behavior—thus aggressively trespassing into the field of linguistics—this argument prompted a strong negative reaction by linguist Noam Chomsky, who wrote a devastating critique of Skinner's book.[11] Other critiques of behaviorism shortly followed and eventually resulted in the demise of behaviorism as the dominant paradigm in psychology. Similarly, when biologist E. O. Wilson became excited about the power of evolutionary explanations of behavior and announced in his book *Sociobiology: The New Synthesis* (1975) that sociobiology was going to take over all the other behavioral disciplines, he did more harm than good to the field: sociobiology was criticized on multiple fronts and eventually evolutionary behavioral scientists stopped using this term so that they could work in peace.[12] In academia there is a lot of territoriality, and threats of intellectual takeovers by new theories or disciplines inevitably elicit strong defensive reactions; some react with an outright rejection and dismissal of the new paradigm, while others are prompted to an extensive scrutiny of the new paradigm, which usually brings to light some of its weaknesses and limitations.

Like many of their illustrious predecessors, in their book *The Handicap Principle: A Missing Piece of Darwin's Puzzle*, Amotz Zahavi and his wife Avishag pushed their theory in all kinds of directions and made many claims about the wide range of phenomena this theory could explain. As a result, the book was met with some skepticism and did not become a best seller. Ultimately, however, the usefulness of a scientific theory is judged by the amount of new research and new knowledge that it generates, regardless of whether or not it lives up to its claims or the extent to which its principles are accepted or rejected. In this regard, the HP has generated a great deal of new research and knowledge and therefore must be considered a very influential and successful theory. Amotz Zahavi's perseverance in presenting and defending the HP, however, incurred the cost of considerable personal and professional strain. In a 2003 article, he made some bitter statements as he noted that living and working in Israel had protected him from the attacks and rejections by his peers, who were mainly based in the United Kingdom and the United States. He wrote: "If I had been dependent on my colleagues for the advancement of my scientific career or my social status, I would not have been able to continue developing the Handicap Principle over the many years in which it was unanimously rejected. Luckily I was living in a corner of the world and . . . at home, my social status and my scientific career were well secured."[13]

Zahavi's ideas about the testing of social bonds represent one of the relatively minor applications of the Handicap Principle. Essentially, while the HP maintains that individuals have to take on handicaps to prove the reliability of their signals, the bond-testing hypothesis suggests that individuals must impose handicaps on others in order to extract reliable information on their attitudes toward themselves. Unlike other aspects of the HP, the bond-testing hypothesis has not been scrutinized and tested with game-theoretical models that verify whether this mechanism is compatible with and can be supported by known evolutionary processes. Researchers have not systematically gathered data to provide conclusive evidence that animals test their bonds with handicap-related mechanisms. Over the years, however, researchers have accumulated a growing number of observations of seemingly paradoxical behaviors in which animals and humans engage in the context of social bonding, and these behaviors appear to be consistent with Zahavi's view.

Eyeball-Poking, Sex, and
Other Bizarre Tests of Social Bonds

Aside from his theoretical work, Amotz Zahavi has observed and studied for many years the behavior of the Arabian babbler, a bird species that lives in Israel. Arabian babblers live in groups of between two and twenty individuals and cooperate in breeding and defending their common territory against neighboring groups. When Zahavi first presented his ideas about bond-testing through the Handicap Principle, he used examples from the species he was most familiar with, the babblers.[14] He described how a male sometimes behaves aggressively toward females during courtship; females who are not interested in the male leave his territory and never return, whereas those who have a genuine interest persevere despite the repeated attacks. In Zahavi's view, this aggression is a handicap that male babblers impose on females to test their suitableness as mates. Zahavi also argued that babblers use allopreening to test the strength of their bonds. One bird preens another, typically a close social companion, pecking at the feathers on his face or body while the recipient remains motionless and facilitates the interaction.

In reality, these are not great examples of bond-testing. Male aggression toward females in the context of mating and allopreening in birds are better accounted for by other explanations. Males in many animal species, including humans, use aggression toward females as a means of social or sexual coercion. Allopreening in birds is similar to allogrooming in primates, and in both cases these behaviors are effectively accounted for by hygienic and social functions. Examples of animal behaviors do exist for which functions other than bond-testing cannot be easily imagined, but descriptions of these behaviors were not yet available at the time Zahavi published his ideas.

For a male baboon, fondling another male's testicles certainly qualifies as an imposition. However, he doesn't need to place his future reproductive career in a partner's hand in order to test the strength of their social bond. Wasting time and being stressed are less dangerous but nevertheless costly activities. Annoyance can involve

185

both, so tolerance of annoyance is a good indicator of how much one values a social relationship.

Capuchin monkeys are small South American primates that live in large groups comprising numerous adult males and females. As with the macaques, baboons, and chimpanzees, capuchin monkeys live in highly competitive societies in which individuals compete for social status through the formation of agonistic alliances. Susan Perry, a primatologist at UCLA who has observed capuchin monkeys in Costa Rica for many years, has reported that these primates periodically test the patience of their favorite social partners—with whom they form aggressive alliances—by subjecting them to all kinds of physically intrusive and annoying behaviors.[15] For example, a young capuchin monkey may walk up to his favorite social partner, stick a finger up his nose, and wait for a reaction. If their relationship is good, nothing will happen, but if the partner has lost some of the initial enthusiasm about the partnership, the annoying monkey will get smacked. Perry noticed that two capuchin monkeys who have a strong social bond sometimes simultaneously insert their fingers up each other's nose and "sit in this pose for up to several minutes with trance-like expressions on their faces, sometimes swaying." Capuchin monkeys also torture their favorite coalition partners by pulling hairs from their face, biting their ears, or sucking their fingers or toes. In a paper published in 2003 in the journal *Current Anthropology*, Perry and her colleagues argued that the function of these interactions is to test the strength of social bonds—a positive response from the recipient, or even the toleration of such behaviors, would be indicative of a good relationship and the willingness to further invest in the bond. When these impositions are tolerated, the two partners groom each other for long periods and continue to form alliances against other monkeys.

At a conference in London in June 2010, Perry showed the audience a video of another bizarre, highly risky, and quite painful bond-testing ritual in capuchin monkeys: two individuals poking each other's eyeballs. I didn't attend the conference or see the video, but the journalist Michael Balter wrote the following description:

One monkey will insert his or her long, sharp, dirty fingernail deep into the eye socket of another monkey, between the eyelid and the eyeball, up to the first knuckle. In videos Perry played for the meeting, the monkeys on the receiving end of the fingernail, typically social allies, could be seen to grimace and bat their eyelids furiously (as did many members of the audience) but did not attempt to remove the finger or otherwise object to the treatment. Indeed, during these eye-poking sessions, which last up to an hour, monkeys insisted on the finger being reinserted if it popped out of the eye socket.[16]

If Zahavi's bond-testing hypothesis does not explain this bizarre behavior, I don't know what other theory could.

Susan Perry's husband, Joe Manson, who is also at UCLA and studies capuchin monkeys with her in Costa Rica, noticed that adult females often touch and briefly hold the infants of females with whom they groom and form aggressive coalitions. Monkey mothers don't

Figure 7.2. *Eyeball-poking ritual between two white-faced capuchin monkeys. Photo courtesy of Dr. Susan Perry.*

like to have their infants handled this way—there is always a chance that the infant might be harmed—and Manson posited that a mother's tolerance of this behavior indicates that she values the bond with the female perpetrator and is willing to provide coalitionary support to her. He suggested that the logic of bond-testing could apply to infants as well: when they climb on the back of a female who is not their mother, they are testing this female's willingness to take care of them, should the need ever arise.[17] As we'll see later, Zahavi argues that human children do this too.

Bond-testing through risky or annoying intimate interactions is prominent in other animal societies as well. What these societies have in common is individuals forming long-term bonds with others, including the formation of coalitions. Spotted hyenas live in complex societies, called clans, that contain up to eighty individuals who together defend a common territory. Like baboon and capuchin monkey groups, hyena clans are structured by linear dominance hierarchies and contain one to several matrilines of adult females and their offspring, as well as multiple adult immigrant males. The spotted hyena is one of the few species of mammals in which females are dominant over males. They are larger than males, are very aggressive, and have an enlarged clitoris that looks like a penis. In spotted hyena societies, females wear the pants, so to speak. Females acquire and maintain their power through the formation of coalitions with other females.

In hyena clans, the commitment problem surfaces in the social bonds that sustain cooperation and coalition formation between two females. Hyena females have been observed frequently testing the strength of their partner's commitment using intimate greeting ceremonies. During these rituals, which last on average twenty seconds but can go on for as long as two or three minutes, two female hyenas stand parallel to one another with their erect penile clitorises, then mutually investigate and sniff each other's genitalia. When these greetings go wrong, they can result in severe genital wounding. As with the baboons, these are risky interactions in which individuals expose a vulnerable body part to others. A team of spotted hyena researchers at Michigan State University, including Jennifer Smith,

Kay Holekamp, and their colleagues, showed that females selectively engage in greetings with their preferred social companions and that these greetings are conducive to the formation of aggressive coalitions against other hyenas of the same clan, joint participation in wars between clans, and cooperative mobbing of lions.[18]

One interesting aspect of greetings in spotted hyenas, capuchin monkeys, baboons, and other animals is that they occur at times when bond-testing is most needed, but also when the probability that they might go wrong is reduced. In the spotted hyenas, females spend hours, or even days, apart from their coalition partners while foraging on their own. The greetings typically occur when the partners are reunited in the clan and allow them to test and update their social relationships. African wild dogs hunt in groups, and efficient cooperation is essential for a successful hunt. Research by Scott Creel and his collaborators has shown that greetings between wild dogs typically occur right before the group is about to go hunting.[19] If greetings occurred during competitive activities such as feeding or mating, someone might get hurt. So the most likely function of greetings is not to reduce tension at particular times—social animals reduce tension in other ways—but to test the relationship. Observations of greetings in domestic dogs made by Barbara Smuts and her collaborators have shown that dogs greet other dogs most often when there are no immediate resources at stake.[20] Thus, greetings appear to offer a mechanism by which animals assess the cooperative tendencies of potential allies in contexts in which the risk of injury is reduced.

Domestic dogs are totally dependent on their owners, and it's important for them to test their bonds to know whether they can sleep safe and sound or whether they should worry about being dumped on the street. According to Zahavi, when your dog jumps in your lap and licks your face or gets in the way of what you are doing, he or she is imposing on you to find out how much you still love him or her and are committed to the relationship. The need to gather information is especially important after the owner has been away for a while or is getting ready to leave; these are crucial moments for finding out where the relationship stands. Of course, you might think that when your dog licks your face, he or she is just showing you affection, but one has to wonder why affection is displayed in that particular way.

Zahavi argues that expressions of love and affection often contain stress-producing, even aggressive elements because the recipient's acceptance and tolerance of them provides reliable evidence of his or her current willingness to continue to invest in the relationship. In this view, many of the affectionate behaviors shown by children toward their parents, such as jumping on their laps or on their backs, derive their communicative value from being inherently stress-producing.[21] Seen in this light, all of our love signals are impositions of one sort or another: kisses, hugs, and petting intrude on personal space and impair freedom of movement. "Lovers who hold each other's hand for hours at a time are each giving up the use of a hand for that time, a pretty heavy imposition," Zahavi says. Lovers who exchange long and passionate kisses stick their tongues in each other's mouth; that's quite intrusive and even carries the risk of transmitting disease. Only people who are highly committed to a romantic relationship accept this type of imposition from their partner. According to Zahavi, bond-testing through costly or stressful love signals is frequent especially when the relationship is new and not yet well established, because this is the time when information is most needed. When partners in a long-term relationship stop exchanging French kisses, it may be that they are not as physically attracted to each other as they were at the beginning, but it also could be that their relationship is strong and bond-testing is no longer necessary.

You will not be surprised to hear, at this point, that Zahavi thinks that sex is the ultimate mechanism for bond-testing. The intrusiveness of many forms of sexual behavior, according to him, makes sex an ideal handicap signal for conveying and receiving detailed information about each lover's commitment to the relationship. Here I respectfully disagree. Although it is certainly true (if unfortunate) that some people find the intimacy of sexual acts a little uncomfortable, the vast majority of people find the intrusiveness of sex quite pleasant and rewarding. And I doubt that many people would regard holding hands as a heavy imposition. The importance of frequently assessing a partner's commitment in a romantic relationship, however, cannot be overestimated. This is what an expert on the subject, evolutionary psychologist David Buss, argues in his book *The Dangerous Passion: Why Jealousy Is as Necessary as Love and Sex*:

> Commitments can change from day to day as a function of an individual's financial situation, or reputation, or age, health, stress, or status. A woman who overestimates her lover's commitment risks abandonment, damage to her reputation, and the hard work of raising a child alone. Overestimating commitment also leads to opportunity costs: the time spent with an undercommitted partner reduces the chance to attract a better mate. Underestimating the true level of a partner's commitment can also be costly, leading to a self-fulfilling prophecy. Miscalculating, for example, could cause you to reduce your own commitment, impelling your partner to do the same, thereby producing a downward spiral of mutual retreat and resentment. The bitter result could dissolve the relationship as both partners search their social world for deeper, more meaningful engagement.[22]

Zahavi's contention that love expressions, including sex, are wasteful, risky, stressful, or even painful impositions—all the properties of handicaps—seems a little pessimistic, but to wonder why romantic partners express their love by sticking their tongues in each other's mouths instead of, say, playing chess together is to ask a legitimate question that, if nothing else, provides food for thought. Zahavi has also made the less controversial suggestion that his bond-testing hypothesis explains why old friends sometimes tease and insult or slap and hit each other.[23] Verbal and physical offenses are clearly impositions that only an old friend who is quite committed would tolerate. In the film *Gran Torino*, the main character, played by Clint Eastwood (who also directs the film), is a grumpy old Korean War veteran named Walt who is trying to teach his young apprentice, a Hmong teenager named Thao, how to navigate social life. Walt is

friends with a barber in his neighborhood, and the two of them exchange banter and racial insults every time they see each other. One
day Walt brings Thao into the barber's shop, and after greeting his
friend with racial slurs as usual, he turns to Thao and says to him:
"You see, kid, that's how guys talk to one another. Now go out, come
back in, and talk to him like a man, like a real man." Thao reluctantly
does as he is told: he walks back into the shop and says to the barber,
"What's up, you old Italian prick?" The barber gets mad at Thao and
threatens to blow his head off with a rifle. The scene nicely illustrates
the principle behind Zahavi's bond-testing hypothesis: one is willing to tolerate an imposition from a good friend but not from a
stranger; thus, imposing on another individual can provide reliable
information about the quality of your relationship.

Can the Commitment Problem Be Solved?

Regardless of whether or not the Handicap Principle is the best explanation for the bizarre behaviors discussed in this chapter, Zahavi's
approach is important because it raises the question of why bonding
rituals—and more generally expressions of affiliation and affection—
are the way they are. One could argue that engaging in any joint activity, whether fondling each other's testicles or standing side by side
and watching the sun set, can represent a bonding experience for
two individuals. Moreover, monitoring your partner's behavior
while engaging in a joint activity can provide some information about
the way he or she feels about the relationship. In reality, however,
the bonding rituals observed in animals are not arbitrary activities,
and the same can be said for some human rituals as well. It is not an
evolutionary accident that so many bonding rituals in animals consist
of risky interactions involving vulnerable body parts or acts that are
otherwise physically intrusive and stressful. And although, in theory,
any aspects of a partner's behavior in a joint activity could provide
hints about his or her commitment, in practice the amount and level
of imposition a partner is willing to tolerate provides more reliable
information than anything else.

The HP applies this approach not only to bonding rituals but more generally to all aspects of animal and human communication. Human language is a peculiar form of communication in which arbitrary sounds or gestures are paired with objects or concepts. Arbitrary words and gestures that have a culturally agreed-upon meaning are referred to as *symbols*. In human nonlinguistic communication and in animal communication, however, the signals are not arbitrary. Rather, they are specifically designed to elicit particular responses in their receivers. For example, the alarm calls given by animals after spotting a predator have acoustic characteristics that are designed to elicit attention and arousal in the listener, while the pain cries of animal and human babies have acoustic characteristics, such as high frequency and high pitch, that are designed to elicit anxiety in the caregivers. Consistent with modern evolutionary approaches to communication, the HP raises the question of why expressions of love and affection in humans are "designed" the way they are (why do lovers kiss the way they do?) and suggests that some of these signals have stress-inducing properties. Although the validity of this and other suggestions remains to be tested, the approach is sound, and the questions being raised are valid. Thus, in addition to bringing economic cost-benefit analyses to the forefront of animal behavior research, the HP has also contributed to the incorporation of modern evolutionary approaches into the study of communication and enhanced our understanding of the design and function of signals.

But going back to the issue of cooperative relationships raised in the previous chapter, does bond-testing with handicaps really solve the commitment problem? Are relationships in which the commitment is frequently tested with impositions more stable and happier than those without such bond-testing?

Leo Tolstoy's *Anna Karenina* famously opens: "Happy families are all alike but every unhappy family is unhappy in its own way." Romantic and marital relationships can come to an end for many different reasons, including, among others, changes in costs and benefits to one or both partners, changes in feelings, accomplishment of goals, chance encounters with other potential partners, and so on.

The same is true for business partnerships and other kinds of cooperative relationships in humans or in animals, especially when these involve individuals that are not genetically related to each other. What the baboons seem to tell us is that when you are in a cooperative relationship, you must keep an eye on it at all times so that you won't be caught by surprise when things change. However, I doubt that testing the bond in and of itself can make a relationship stronger or more stable, unless it is accompanied by other efforts. Anyone who has seen soap operas or *Oprah* knows what these efforts are: keep your romantic feelings alive; make sure that the cost-benefit ratio is always favorable for both of you; make sure you always have clear and important shared goals; and don't look around for other options lest you fall prey to temptation.

Chapter 8

Shopping for Partners in the Biological Market

Finding the Right Partner

Cooperative relationships—whether it's an agonistic alliance between primate males, a business partnership between restaurant co-owners, or a marriage between lovers—all have one common risk: things can go wrong because one partner cooperates one moment and cheats in another, or one partner promises eternal commitment one day and ends the relationship the next. So before we enter into a cooperative relationship we should do some research about how our partner has behaved in the past, and then monitor his or her every move; play tit-for-tat; provide incentives and rewards for cooperation as well as discouragement and punishment for cheating; use feelings, morality, religion, and the legal system to make our partner behave; and finally, check the strength of his or her commitment with bizarre, risky, annoying, or sexually daring behaviors on a daily basis.

Despite all of these precautions, a relationship between two individuals can still go wrong. One simple reason for this failure may be that we picked the wrong partner to begin with—someone who is not a good cooperator in general, or not a good match for us. The success of our cooperative relationships may depend in large part, not on how we or our partners behave, but on who we choose as our partner and whether or not he or she is the right person for us.

Economists and evolutionary biologists have developed two different types of theoretical models of cooperation—those that focus

195

on partner control and those that focus on partner choice. Models of partner control, such as the Prisoner's Dilemma, take the formation of a cooperative partnership as a given and concentrate on the strategies that each partner uses to avoid being cheated by the other. Successful strategies in the Prisoner's Dilemma are those based on the partner's behavior in the past and the probability of cheating in the future. In this and other models involving one pair of players, it is assumed either that the pairs are formed by some external entity—say, the police officers who arrest and interrogate two suspects of a crime—or that individuals are paired off randomly. For each player, the only alternative to an interaction with the partner is no interaction at all. In the real world, however, people often choose a partner out of many individuals. And in nature, situations in which individuals choose their partners are generally more common than those in which partners arise arbitrarily or are randomly assigned. When animals do business with someone else, first they choose their partner very carefully—after sampling many potential candidates for the job, either at the same time or one after another—and then they monitor and control their partner's behavior to ensure continuing profit from their venture.

As I discussed in Chapter 6, the process by which employer and employee or landlord and tenant select each other shares many features with the process through which people search for a romantic partner. In both kinds of cases, individuals operate in a market regulated by the laws of supply and demand. Evolutionary biologists have taken this a step further: they have shown that the process of partner choice in the human employment and marriage markets is remarkably similar to the process through which all organisms—including viruses, bacteria, plants, and animals—find partners for all kinds of cooperative social relationships.[1] Some of these enterprises include mutually beneficial interactions between different organisms, such as plants and insects, parasites and their hosts, or different animal species such as the cleaner fish and their clients (discussed in Chapter 5). The same models developed by economists and evolutionary biologists can explain partner choice in all of these seemingly disparate aspects of life. We'll begin with something everyone is familiar with: the human mating market.

Human Mating Markets

Walking on the streets of Bangkok a few years ago, I couldn't help but notice the high number of heterosexual "mixed" couples made up of a white Caucasian man and a Thai woman. In virtually all cases, the man was older and rather unattractive (bald, with a potbelly and thick glasses), while the woman was young and good-looking. We see well-matched couples all the time: the young and the beautiful typically go with their own kind (like Brad and Angelina), and average-looking middle-aged people are typically married to other average-looking middle-aged people. Occasionally, we run into a very attractive young woman accompanied by an older man, but the man is typically well groomed, in good physical shape, and wearing an expensive Armani suit. In other words, he is wealthy and success-ful. In the United States or in Europe, you don't typically see unat-tractive and socially awkward middle-class men in the company of beautiful young women.

So what's going on in Bangkok? Why do we see all these pairings that seem mismatched for age and looks, and all in the same direction? Why don't we see young and handsome Caucasian men dating older and average-looking Thai women? The Thai women I saw were not escorts—men and their female escorts usually don't walk around in the middle of the day holding hands. It's much more likely that these couples were dating or even engaged. These men travel to Bangkok from the United States or Europe, meet these beautiful women, marry them, and bring them back to their country. After returning from Bangkok, I started noticing similar mixed couples in the United States—with the same differences in age and looks—but generally older. These were the couples who probably met in Bangkok ten or twenty years ago and have been living together in the States ever since.

I am sure that anthropologists and sociologists have many good ex-planations for this phenomenon, but so do economists and evolutionary biologists. The latter believe that there is a mating or marriage market in which individuals have characteristics that make them more or less attractive to members of the other sex and in which partner selection is regulated by the laws of supply and demand. Individuals who have low value and little bargaining power in one market can move into another, where their characteristics are more in demand. Let me explain the way a mating market works in a little more detail.

Everyone in the mating market has certain endowments that others may find attractive, such as youth, physical attractiveness, wealth, and social status. Age and physical traits, in particular, are what people often consider first in a potential partner. It's like looking for ripe cantaloupes at the fruit stand—there are hundreds on display, but you won't even touch any that are too small or still green. You pick up the ones that meet your criteria for size and color and start palpating them, looking for other indicators of ripeness. Likewise, when a partner has met your initial criteria of age and attractiveness, you consider other traits such as status, wealth, intelligence, honesty, or generosity. That physical characteristics are evaluated first has been shown by many studies conducted by psychologists, including some involving speed-dating, which I discuss later.

Males and females differ, on average, in how they value the endowments of opposite-sex individuals. Men value youth and physical attractiveness very highly, while women value wealth and status (though they don't mind physical attractiveness too). Clearly there are not enough young, beautiful women for every man, so a few men get them and most don't. On the flip side, since young beautiful women are in short supply and high demand, they can select any partner they want. Angelina Jolie found a partner who has all the characteristics that arguably every woman wants: Brad Pitt is (relatively) young, handsome, wealthy, healthy, famous, and powerful, and he also seems like a nice guy and a good father. Well-endowed men with good looks, lots of money, and high social or celebrity status are also rare and in high demand, so they too usually get what they want. But men with low endowments, such as low income or average looks—and there are many of them out there—have limited options.

If they are nice and have good social skills, they can settle with a partner with similarly low endowments, but if they happen to be socially awkward or unpleasant to be around, they may not find anyone at all.

However, in the era of globalization, when it's easy to travel around the world and meet people through the Internet, the low-endowment men have another option: move into a different mating market, one in which their endowments are deemed more valuable. In Bangkok, where local people are quite poor, a middle-class American man is considered a billionaire. Most of all, marrying an American man offers a Thai woman the opportunity to climb out of poverty, leave her country, become a U.S. citizen, and maybe spend the rest of her life in a suburban house in Florida or California. Thus, in the Bangkok mating market, middle-class, middle-aged American men, despite their baldness, potbellies, and thick glasses, are considered by many Thai women to be more valuable as potential husbands than most Thai men. In this market, American men can pick and choose, and of course they all want young, beautiful women.

Of course, this is an oversimplification of how human mating markets work. What's considered valuable in a partner varies depending on whether one is looking for a short-term, mostly sexual relationship or a stable long-term relationship involving marriage and children. Women's preferences for men's traits vary depending on what stage of their menstrual cycle they are in: they value good looks and masculinity more around midcycle than at other times.[2] Finally, there are cultural differences in partner valuation. People living in Manhattan may have a different idea of what is attractive in a potential partner than people living in rural villages in New Guinea.

The general point is that when people "shop" for a partner in a mating market, they can't always get what they want. What they can get depends on their own value and on the laws of supply and demand in the particular market in which they find themselves. The fact that most women and most men seem to value the same qualities in a partner does not contradict the observation that people often end up in relationships with individuals who were not at the top of their wish list. We would all like to live in big mansions, but in practice people live in houses they can afford. Similarly, although people generally agree on who are the most desirable mates, they end up with someone whose value is comparable with their own. Knowing whether one is a 2, a 6, or a 9 on the 1–10 scale of mate value is important, and that's not something one can figure out by looking in the mirror. It takes time as well as feedback from our fellow human beings.

When adolescent boys and girls first enter the mating market, experimentation with dating allows them to assess their own mate value: some adolescents discover that they possess traits that are in high demand and make them popular and successful, so they become very choosy; others experience indifference or rejection and realize that they need to be content with the low value of their endowments or work hard at increasing it. According to evolutionary psychologist David Buss, author of *The Evolution of Desire: Strategies of Human Mating*, it is important that anyone reentering the mating market later in life—for example, after a long marriage that ended in divorce—reassess his or her value on the current mating market:

> The existence of children from the previous marriage generally lowers the desirability of divorced people. The elevated status that comes with being more advanced in their career, on the other hand, may raise their desirability. Precisely how all these changed circumstances affect a particular person is often best evaluated by brief affairs, which allow a person to gauge more precisely how desirable he or she currently is, and hence to decide how to direct his or her mating efforts.[3]

The notion that there are mating or marriage markets, of course, is nothing new. Economist Gary Becker at the University of Chicago conducted this type of analysis almost fifty years ago.[4] Similar studies have been carried out more recently, such as the work by economist Shoshana Grossbard-Shechtman that resulted in the 1993 book *On the Economics of Marriage*.[5] Evolutionary psychologists have made their contributions to our understanding of human mating markets by studying people's personal ads or their preferences in speed-dating encounters.

Personal ads published in newspapers or posted on dating sites such as Match.com serve to advertise both the poster's own characteristics as well as his or her needs: in other words, their offers and their demands. Ads can be viewed as bids that reflect both people's self-assessment of value and their knowledge of the market. Various studies of ads have shown that advertisers adjust their bids in light of their perceived market value. In highly competitive markets, individuals with a weak bargaining hand adjust their demands down, while those with a strong bargaining hand adjust them up. In 1999, evolutionary psychologists Boguslav Pawlowski and Robin Dunbar conducted a study of ads in which they objectively assessed the market value for particular age and sex classes of individuals.[6] They did this by dividing the proportion of male and female advertisers seeking individuals of a given age (the demand) by the proportion of male and female advertisers of that age (the supply). As expected, they discovered that for women, market value peaks in their late twenties, while for men it peaks in their late thirties. Accordingly, ads posted by women and men in these age categories were those that received the most replies. They also found that women and men with high market value were more demanding and choosy, looking for many specific traits in a potential partner.

Other interesting findings come from studies of speed-dating. HurryDate is a speed-dating and online dating company for adult singles living in major metropolitan areas in the United States. The people who run this organization set up meeting sessions in which twenty-five men and twenty-five women who have never met before interact with each other for three minutes and subsequently indicate which of the people they met they would like to see again in the future; if there is a match, the organizers give email addresses to the individuals so that they can get in touch directly and arrange a more traditional date.

Two evolutionary psychologists, Robert Kurzban and Jason Weeden, analyzed behavioral and questionnaire data from 10,526 participants in HurryDate sessions.[7] They discovered that among both men and women, some individuals are in high demand (their mate value is high) and others are not. The characteristics associated with high mate value in this situation were almost exclusively physically observable attributes such as attractiveness, slenderness, height, and youth, whereas harder-to-observe attributes such as education, religion, sociosexuality, or views on children didn't matter at all.

Consistent with the idea that people are aware of their participation in a market and know how to navigate it, high-value individuals were picky and ended up selecting only individuals who had high mate value as well. Conversely, less desirable men and women were also less selective about their partners. For example, heavier women agreed to a relatively high proportion of potential dates, as did men who were either heavy or very thin. In this study, desirability was directly related to physical attractiveness for both men and women. Clearly, given the logistics of speed-dating, it is not surprising that in this context, physical attractiveness is the most important factor for partner selection. Other studies investigating online dating, however, have found that the best predictors of the number of opposite-sex emails received is physical attractiveness in the profile photo for women and income for men. Good-looking women and high-income men get the most emails.

Biological Markets

In nature, situations in which individuals cooperate with one another can be thought of as markets in which commodities of similar or different value are exchanged according to the laws of supply and demand, usually through advertisement and bartering. We call these situations *biological markets* to distinguish them from human markets in which money is used as currency. Many biological markets involve two distinct classes of traders: in the mating and reproduction market the traders are males and females, and in the alliance market they are high-ranking and low-ranking individuals. Traders can also be animals of different species, such as cleaner fish and their customers, or plants that need to be pollinated by insects and the insects that pollinate them.

In the mating market, males offer the following commodities to females: sperm to fertilize the females' eggs, good-quality genetic material that will make offspring healthy and attractive, and help with child-rearing. In some animal species, males also offer territories that contain food, nests in which females can lay their eggs, or so-called nuptial gifts involving food that can be consumed by females before or during mating. Females, in turn, offer eggs to be fertilized by males, bodies in which the embryos can grow, and the ability to take care of the babies. In primate markets, as we'll see later, monkeys can trade grooming for other grooming, for sex, or for agonistic support.

In biological markets, as in other kinds of markets, some traders possess commodities of higher value or quality than others do. In the case of territorial animals, for example, some males may have larger territories, or territories containing more food or better nests, than other males. Traders of one class can choose among traders of the other class in relation to the value of their commodity, but they must compete with members of their own class for access to preferred potential partners. Males can choose a mating partner among many different females, but must also compete with other males to mate with highly attractive females. Partner choice is important in a market because doing business with selected individuals who possess high-value commodities is generally more profitable (for animals, *profit* means increased survival and future reproduction) than doing business with a random individual. I emphasize the term *choice* because biological market theory assumes that commodities cannot be obtained through force but only with the consent of the trading partner. Similarly, competition with members of one's own class does not generally involve aggression or intimidation. For example, males who compete with each other to mate with an attractive female cannot eliminate each other by force. Rather, traders in the same class compete by trying to outbid each other in the value of the commodity offered: they try to offer better products than their competitors. In turn, decisions to cooperate with a particular partner—partner choice—are based on the comparisons between the offers of different potential partners.

The issue of comparing the offers of different bidders is of crucial importance in biological markets. When a female chooses to mate with a particular male on the basis of the territory he possesses, she must have the ability to directly assess the quality of the territory, to inspect the territories of different males, and to make comparisons among them. The sampling of individuals with different commodities can be a complicated, expensive, and time-consuming process. It is often simply too expensive to sample all bidders, and one can sample only a subset of prospective partners. Thus, when evolutionary biologists develop market models to predict the partner choice decisions made by particular individuals in particular situations, the costs of sampling and assessment are crucial variables because they determine whether the number of potential partners is high (when costs are low) or low (when costs are high).

The accuracy of the strategy and procedures used to sample potential bidders and assess the quality of their commodities is also an important variable. In some biological markets, the value of the commodity can be assessed directly. In insect species in which males bring females a nuptial gift, the female can immediately assess the size and quality of the gift. In other markets, however, assessment is based on an advertisement that is supposed to represent the quality of a commodity. Male birds often advertise their good health, strength, and high social status through brightly colored stripes or spots on their plumage. Where there is advertisement, however, there is also potential for false information. Just as people who watch TV commercials cannot tell whether the qualities of the product being advertised are real, when females assess the quality of potential mating partners indirectly, through signals produced by the males themselves, they can't be sure the males are being honest. Thus, among traders there may be individuals who pretend to offer a particular high-quality commodity but then can't deliver on their promise—so-called free riders. I already discussed the issue of honest and deceitful signals when I mentioned the Handicap Principle in Chapter 7. The use of signals to advertise commodities implies that when traders do business with each other, they not only inspect and compare the commodities of different potential partners but also communicate directly with them, engaging in bartering and bargaining.

The exchange value of a commodity in a biological market is determined by the ratio between supply and demand, and this can vary over time. As we'll see later, the value of a nest as a commodity that male birds offer to attract females may vary during the course of a year depending on how easy or difficult it is to build nests, how many males are able to do it at a particular point in time, and how urgently females need these nests to lay their eggs. Studies of biological markets have shown that the establishment of cooperative relationships between individuals, including partner choice, changes over time in relation to temporal fluctuations in the supply and demand of particular commodities.

Another important characteristic of biological markets is that they are often skewed, which means that one commodity is in high demand and another one isn't. This may happen because the members of one trading class outnumber the others, or simply because one commodity is abundant and another one is rare. Women's eggs are in shorter supply than men's sperm because women have only one mature egg per month until they reach menopause, while men can produce millions of sperm every day of their lives. Traders who possess a commodity that is high in demand become the choosing class and can find a partner easily, while traders with a commodity that is low in demand become the chosen class and have to outbid their class members to find partners. In the mating market, females can contribute to this process of competition by playing males off against one another and forcing them to increase their bids during courtship, such as by offering food or services or engaging in risky behavior— or in the case of humans, simply spending a lot of money. Traders who can't afford to make competitive bids are forced to take less favorable options and settle with less valuable partners.

Let me now provide some examples from the animal kingdom that illustrate some of these general principles regulating "business" in biological markets.

ANIMAL MATING MARKETS

Primatologist Michael Gumert examined the mating market in wild long-tail macaques, a species similar to rhesus macaques that lives in the forests of Indonesia.[8] These monkeys live in large groups with many females and many males. Females are fertile during four or five days in the middle of their menstrual cycle and advertise their estrus with sexual swellings. Normally, at any point in time, half the females in the group are pregnant or breast-feeding a young baby and so are neither fertile nor interested in sex. The menstrual cycles of the other females are usually not synchronized, so they don't all become fertile at the same time. This means that when one female in the group is in estrus, her fertility is a valuable commodity that every male in the group wants. Males cannot sexually coerce the fertile female and cannot use force to prevent their male competitors from mating with her. Instead, they have to offer the fertile female another commodity and outbid their competitors to make sure that she will do business with them. This commodity is grooming. Receiving grooming increases one's hygiene and reduces tension. It's sort of like when a husband offers a back rub to his wife in hopes that she will consent to have sex. That macaques consider grooming a valuable commodity is suggested by studies showing that individuals often mutually groom each other and do so in a time-matched manner, and that when subordinates groom high-ranking individuals they obtain tolerance and support in return. Does grooming also work as payment for sex?

Gumert observed that when there is a fertile female in the group, the males groom her much more than they groom the nonfertile females. He also observed that after a male grooms the fertile female, they often have sex. It turned out that male grooming before sex lasts longer than grooming not followed by sex, and that a female is more likely to have sex with a male after this male has groomed her for a while than if he just sits there and does nothing. So it seemed that being groomed by a male for a while puts the female in the right mood for sex. One interpretation of this is that males use grooming to pay fertile females for sex. In contrast, fertile females don't make any grooming payments to males in order to have sex with them. All fertile females have to do to get some romantic attention from males is to look pretty with their sexual swellings.

Gumert also noticed that not all males pay the same grooming price for sex, and not all females receive the same compensation for their availability. High-ranking males are generally more attractive than low-ranking ones as mating partners because they can provide better protection from other monkeys to females and their offspring. High-ranking females are more attractive as mating partners because they are generally healthier and more fertile than low-ranking ones. As would be expected in a mating market, the value of individuals influences the commodities they are able to obtain as well as the price they pay for them. High-ranking males groom fertile females less but mate with them more than low-ranking males do. Given that low-ranking males are less attractive, they have to work harder at gaining their favors. High-ranking females mate with high-ranking males more often and receive higher grooming payments from males for the same amount of sex as low-ranking females do. Finally, in accordance with the law of supply and demand, if there is only one fertile female and many males in the group at a particular time, the grooming payment made by the guy who gets lucky is very high. If there are more fertile females at the same time in the group, the grooming payment received by each fertile female is lower.

As you can imagine, Gumert's article, aptly titled "Payment for Sex in a Macaque Mating Market," made a big splash with the media when it came out in 2007. Newspapers, magazines, and Internet news sites played around with headlines that more or less explicitly hinted at the discovery of monkey prostitution. Spike TV sent a crew to my office to interview me about this article, but I totally "choked" in front of the camera and I don't think the interview ever aired.

Mating markets are common in birds. A good example is the mating market of the red bishop, a colonial weaverbird living in southern

Africa, which has been studied by German biologist Markus Metz and his colleagues.[9] Red bishop males mate with many different females, which they try to attract by building several nests within their territories. Unlike the many other bird species in which males and females take care of their chicks together, red bishop females incubate eggs and feed the chicks entirely on their own. Females, however, don't build nests; nests are the commodities offered by males when they try to entice females to do business with them.

Females sample many males before choosing a mating partner, and they base their choice entirely on the quality of the nests. Some males build many nests, others only a few; some nests are good, others just okay. The males have adjacent territories, so it's easy for females to hop around from one territory to the next to examine the merchandise; in other words, the costs of sampling are low, and females can afford to examine many nests before choosing one. There are always more nests on the market than females who are looking for them, and as is always the case when the supply is greater than the demand, females are very choosy about nests, while males try to outbid each other by offering better nests than their neighbors. When the female demand for nests increases at particular times of the year—because greater availability of food allows for more intense breeding activity—males build nests more quickly. Since females prefer fresh new nests that have been built within the past week over those that have been on the market for a long time, all males are under pressure to put new nests out on the market.

Choosy females inspect many different nests carefully before settling on a brand-new one in which they lay their eggs. Females become more likely to accept older nests when their own market value decreases, that is, when there are many females searching for nests but only a few nests are available. Therefore, partner choice in the red bishop mating market is regulated by the value of nests and other commodities traded by males and females, which in turn depends on temporal fluctuations in their supply and demand.

Mating markets similar to that of the red bishop are also found in insects, but instead of nests, males offer females food to convince them to mate.[10] In scorpion flies, males offer females smaller insects as nuptial gifts, and females choose their mating partners not only by allowing some males to mate with them but by refusing or interrupting mating with others. Cheating by males is difficult since females can assess the size and quality of the gift immediately. The larger the gift, the higher the probability of successful mating. Studies of scorpion flies have shown a clear market effect on male-female business transactions: females reject males with small gifts when many males offering food are available, but they accept gifts of any size when there are only a few offers on the market. In some species of spiders, males offer themselves as food to females—the ultimate sacrifice—in order to mate with them. When a female accepts a particular male as a mating partner— which means that she is either horny or hungry, or both—she will often start chewing on the male's head while the lower part of his body is still busy copulating. Although I don't know of any market studies of this interesting system in which trading of commodities involves male suicide and female cannibalism, I would anticipate that fat males do better business than thin ones, unless females haven't had a meal in a while, in which case they probably mate with, and quickly devour, males of any size that happen to come their way.

MARKETS FOR BUSINESS OTHER THAN SEX

Believe it or not, not all animal business revolves around sex. Animals also trade in other commodities. In monkeys, grooming can be used as payment for sex, but it can also be traded for grooming itself or for other goods or services. In many cases, low-ranking individuals offer grooming to high-ranking ones in exchange for being left in peace while eating, or for protection when they are under attack by their monkey enemies. Childless females also offer grooming to mothers to be able to take a close look at, touch, and briefly hold their baby, a highly valued commodity among primate females. Market effects have been demonstrated in all of these contexts. Primates make higher grooming payments when commodities are in short supply: low-ranking monkeys groom high-ranking ones for longer periods of time in order to be tolerated around food during periods of food shortage when compared to periods of food abundance; and females groom mothers for longer periods to be allowed to touch their infants when there are fewer infants in the group, compared to when there are many infants.

Let's take a closer look at the market in which grooming is traded for tolerance, protection, or support. In primates living in large groups, such as macaques, baboons, and vervet monkeys, social interactions within the group are regulated by nepotism and dominance. Family members associate with one another and exchange grooming more often than they do with nonkin. While monkeys groom their family members because they "love" and support them, grooming between nonkin generally has a business purpose. In this regard, high-ranking and low-ranking individuals can be thought of as two classes of traders that exchange commodities. High-rankers can offer low-rankers tolerance around food, protection from harassment by group members, and agonistic support in fights against others. Therefore, high-rankers are valuable and attractive social partners with which to do monkey business. Low-rankers are tolerated within a group probably because they help out when the group has to fight against predators or other groups. Low-rankers fight in the front lines during wars with other groups, and if a predator attacks a group, they are more likely to be eaten first. When the group is at peace, however, and low-rankers don't have to risk their lives on behalf of everyone else, the only commodity they can offer high-ranking individuals in exchange for their tolerance and help is grooming.

Of course, there is variation among high-rankers in the quality of the services they can provide to subordinates; the higher their rank, the more valuable their service. Among females, the alpha female is the most valuable partner. Low-ranking females compete with and try to outbid one another to obtain the services of their queen. They do so by trying to groom the alpha female whenever they get a chance, and for as long as possible. The value of grooming as a commodity increases with its amount, so that the more grooming the alpha female receives from a particular female, the more she should be willing to tolerate and protect that female.

Compared to the two-sided mating market in which both males and females select each other as partners, the monkey grooming market is rather one-sided, because, to a high-ranking female, the benefits of receiving grooming are pretty much the same regardless of who does the grooming. All low-rankers are the same—it's just labor. In addition, since there is usually a large supply of low-ranking females in the group eager to groom all day, high-rankers don't have to compete with one another to obtain their services. Even high-rankers, however, have their own preferences for social partners. They prefer to hang out with kin, for example, rather than with nonkin. So if the daughter of the alpha female and the lowest-ranking female simultaneously offer to groom the alpha female, there is a good chance that the alpha female will accept her daughter's offer and refuse the other's. The alpha female's daughter ranks just below her mother in the hierarchy, and in general kin are closer to each other in rank than nonkin. This means that, in the competition to groom the alpha female, the bargaining power of low-ranking females decreases as their distance in rank from the alpha increases. Biological market theory predicts that individuals in weak bargaining positions will become less selective and lower their demands. Thus, even though every female in the group would be happy to give all of her grooming to the alpha female, in reality the probability of this happening decreases the lower the female's rank.

In the 1970s, long before biological market theory was developed, primatologist Robert Seyfarth speculated that competition between low-ranking females to groom the alpha female and the constraints imposed by supply and demand should result in every female having to compromise and groom the female that ranks just above her in the hierarchy. His observations of a group of baboons confirmed his intuitions: while most female grooming was directed up the hierarchy, each female groomed the female one notch above her in the hierarchy most often.[11] These observations have been replicated many times, not only in other primates such as macaques and vervet monkeys but also in other animals, such as the spotted hyenas I discussed in Chapter 7.

Spotted hyenas have a social structure similar to that of baboons, macaques, and vervet monkeys. Even though they don't groom one another, they express their social preferences by joining subgroups that contain particular individuals. Jennifer Smith, Kay Holekamp, and their colleagues at Michigan State University studied the social preferences of spotted hyenas from the perspective of biological market theory.[12] They concluded that although the highest-ranking hyenas in the clan could offer more goods and better services to subordinates, market forces lead subordinates to associate most closely with the animals that hold a rank immediately above them rather than with the highest-ranking females in the clan. Again, this is the result of supply and demand and of competition among low-ranking individuals. But what happens if the market value of an individual as a social partner changes suddenly and dramatically? How is the market affected by this change?

To address this question, a group of researchers led by Dutch primatologist Ronald Noë—a pioneer in the development of biological market theory and its application to animal behavior—conducted an ingenious experiment with wild vervet monkeys in South Africa.[13] After creating artificial markets in two groups of these monkeys by making certain individuals valuable as cooperation partners, the researchers recorded how much grooming other group members were willing to pay them to benefit from their services. Then they experimentally changed the market value of these individuals and looked at what happened to the grooming trade. But let me describe what the researchers did step by step.

At the beginning of their study, Noë and his colleagues simply recorded who groomed whom and for how long in order to show that, as is always the case in vervet monkeys, high-ranking individuals receive a lot more grooming than low-ranking ones. Low-ranking females, unless they are in estrus or have a newborn infant, are not attractive social partners because they have no power and therefore have low value as potential coalition partners. Then the researchers taught a low-ranking female in each of the two groups to press a lever that opened a container full of food. The container contained enough pieces of apple for every group member to have a good chance of getting some apple to eat, even though, as usual, the high-rankers got more than the others. The opening of the container was repeated sixteen times over a period of nine weeks (phase 1). During this period, the researchers recorded all grooming interactions involving the food-delivering female and other monkeys in the hour after she opened the food container.

In phase 2 of the experiment, Noë and his colleagues trained another low-ranking female within each group to open a second food container. The same amount of food was now divided between the two containers, which were made available simultaneously. The researchers again recorded all grooming exchanges between individuals after the introduction of the second skilled individual. They were especially interested in the ratio of grooming given relative to grooming received by the two food providers because they expected that other monkeys would behave nicely to them and offer grooming without demanding any in return. It turned out that before phase 1 of the experiment, the first low-ranking female that was subsequently trained did a lot of grooming but received little in return. During the period in which she opened the food container for everyone, this changed dramatically: she became popular and received a lot of grooming while giving very little. Interestingly, she opened the food container preferentially in the presence of those who groomed her the most, thus giving them the chance to grab a lot of apple pieces. When the other monkey started opening the second food container, however, the market value of the first food provider dropped by half: she was still getting more grooming than she had initially received, but the effect was half as strong as before. These changes were observed in both groups. Thus, the trading of commodities in primate grooming markets varies in relation to changes in the value of the individuals as trading partners, exactly as predicted by biological market theory.

Trading Between Species: Mutualistic Markets

Mutualism is a cooperative relationship between two organisms of different species in which both benefit from their association. Unlike cooperation, in which altruistic acts are reciprocated with a time delay, in mutualism the same interaction simultaneously benefits both partners. A wide range of mutualistic relationships, not only between animals but also between animals and plants, have been studied from the perspective of biological markets. To illustrate the approach, I first describe the business between ants and the larvae of Lycaenid butterflies, and then the cleaner-client fish market.

THE ANTS-BUTTERFLY LARVAE MARKET

Many species of ants protect Lycaenid butterfly larvae against predators and parasites. In exchange, the larvae offer the ants a sugar-rich nectar, which is produced by a gland called the nectar organ. The only function of the nectar is to attract ants and reward them for their protection. Researchers have discovered that the butterfly larvae adjust the amount of nectar offered depending on the number of ants protecting them. When only a few ants are present, the larvae produce more nectar to attract additional ants; when many ants are present, they reduce their nectar production. Thus, this appears to be a biological market in which the larvae compete with one another to attract ants and adjust their supply of nectar in relation to the demand. When ants are few, intense competition between larvae results in increased bidding and nectar production; when there are many ants on the market, the larvae can afford to reduce their bids and produce less nectar (which is costly to produce). In an interesting twist, larvae that do not produce nectar (free riders that try to benefit from the ants' protection without providing anything in return) sometimes end up being eaten by the ants: if a larva produces little or no nectar, her body becomes valuable as food to the ants. So, by eating these unproductive larvae, the ants kill two birds with one stone: they get their food, even though it tastes less sweet than nectar, and they also eliminate the free riders from the population.[14]

THE CLEANER-CLIENT FISH MARKET

There is a lot more to the story of cleaner fish and their clients than what I briefly covered in Chapter 5. Research conducted by biologist Redouan Bshary and his colleagues has shown that the mutualistic interactions between cleaners and their clients, like those of ants and butterfly larvae, are regulated by the laws of the market.[15] To recap, cleaners are small fish called bluestreak wrasses (their scientific name is *Labroides dimidiatus*) that inspect the body surface and the inside of the gill chambers and mouths of larger fish—the clients— in search of skin parasites and dead or infected tissue. As I mentioned earlier, cleaners sometimes "cheat" and also eat mucus, scales, and fleshy tissues inside the mouths of their clients. These fish are found in the Red Sea and in the Pacific and Indian Oceans, from the east coast of Africa all the way to the Great Barrier Reef in northeastern Australia. Bshary and his colleagues observed these fish at Ras Mohammed National Park in Egypt and did laboratory experiments with them at the Lizard Island research station at Australia's Great Barrier Reef.

The cleaners live in small territories, called cleaning stations. Clients visit these cleaning stations and often use special postures (such as spreading their pectoral fins and stopping their swimming, maintaining a head-up or head-down posture) to signal their desire to be serviced. Individual clients visit cleaners and seek inspection from them five to thirty times per day, sometimes up to one hundred times per day. Cleaners may perform more than two thousand inspections per day. Clients can discriminate among different cleaners directly, or indirectly, depending on the location of the cleaning station where they meet them. Cleaners seem to be able to recognize their clients just from the way they look.

The cleaner-client system can be thought of as a market in which two classes of traders offer different commodities: hygiene in exchange for food. Cleaners usually stay in their territory; in other words, they stand behind the counter in their shop waiting for the clients to arrive. Clients can decide whether or not to visit a particular cleaning station. Cleaners, in turn, can accept the request for cleaning or ignore the client. There is competition among clients: they often form lines at cleaning stations, waiting to be served. Cleaners have clients from the immediate neighborhood as well as from the open sea. The "resident" clients never leave the neighborhood in which

Figure 8.1. The cleaner wrasse at work with one of its clients.

they live and therefore have access to only one cleaning station, while "floater" clients normally swim over larger areas that include several cleaning stations. The cleaners have exclusive access to resident clients in their territory without competition from other cleaners: they are the choosing class. The resident clients are the chosen class, which means that sometimes they have no choice but to accept poor service because poor service is better than no service at all; they wait longer, get a shorter cleaning, and occasionally have to put up with being bitten by their cleaners. In contrast, the floaters have access to different cleaning stations, so they can select the cleaner that gives the best service. As clients, the floaters can afford to be choosy, so cleaners try to outbid each other by providing better service to attract them.

Two floater clients or a resident and a floater can compete for services from the same cleaner. Sometimes a client arrives at a cleaning station while the cleaner is inspecting another client; other times, two or more clients seek inspection from the same cleaner simultaneously and the cleaner has to choose between them. Competition between clients occurs only through invitations to the cleaner for inspection, not through aggression. When a cleaner chooses between clients, biological market theory predicts that it will choose the more valuable client, the floater, over the resident, because if the cleaner ignores the floater it will lose it as a client, possibly for good, whereas the resident will continue visiting the shop—it has nowhere else to go—even if it is temporarily ignored. So if a resident needs cleaning but a floater has already occupied the station, the resident has no option but to wait in line for service or to come back later. In contrast, if a floater doesn't get immediate attention from a cleaner, it can go to another station and never come back.

Observations and experimental results show that cleaners do indeed give preferential treatment to the floaters over the resident clients. In one study, cleaners switched from a resident to a floater client fifty-one times, but switched only once from a floater to a resident. When a floater and a resident client requested service simultaneously, the cleaners inspected the choosy client in sixty-five out of sixty-six instances. Finally, floaters got better service: they got cleaned early in the morning (a favorite time for all client fish, since nothing is better than a shower before going to work); they never waited in line; they were cleaned for longer durations; and most importantly, they didn't get bitten by a cleaner. When a floater was occasionally ignored or bitten by a careless cleaner, he immediately went to a different station and was never seen again in the shop.

It's interesting that the cleaners prefer the floater over the resident clients even though sometimes the residents are larger and presumably have more parasites (which means more food for the cleaners). Bshary did some very careful experiments with fake floaters and residents to make sure that cleaners didn't always simply choose the "fat" clients. When cleaners have to choose between two floaters, however, they typically choose the larger one, which has more parasites as well as more mucus they can chew on. Because Bshary always examined cleaners' preferences for clients of different species, however, it is possible that other differences between these species accounted for his results. This possibility was addressed by Thomas Adam, a behavioral ecologist at the University of California–Santa Barbara, who recently published an article with the clever title "Competition Encourages Cooperation: Client Fish Receive Higher-Quality Service When Cleaner Fish Compete."[16] Instead of examining cleaners' partner choice between clients of different species, Adam focused on a single client species, the ornate butterflyfish, *Chaetodon ornatissimus*. Some individuals of this species have small territories containing only one cleaning station, and others have larger territories with multiple cleaning stations. Multiple cleaning stations means competition for the cleaners; market theory predicts that cleaners should provide higher-quality service to clients with multiple cleaner stations in their territory. Adam's study showed just that. The clients with more options were cleaned more quickly and for longer periods of time, while clients without options were forced to wait in line.

Back to Bshary's work. There is an interesting twist in his story of the cleaner-client fish market. It turns out that, although the majority of client fish are vegetarian and only eat algae, about 15 percent of client species are carnivorous and eat other fish. This means that when cleaners get into the mouths of these predators, they risk being swallowed and digested, especially if they don't do a good cleaning job and inadvertently (or intentionally) hurt the client. When a cleaner cheats with a vegetarian client, the client simply "gets mad" and swims off or chases the cleaner, but when a predatory client gets mad, the consequences can be far more serious. In terms of market dynamics, the cleaners simply trade hygiene for food with harmless clients, but with the predators they also trade their safety. Consequently, biological market theory predicts that the predatory clients— whether floaters or residents—should receive better treatment than the harmless clients. And lo and behold, it turns out that predators are bitten less often than harmless clients. As mentioned in Chapter 5, a cleaner's biting (cheating) makes a client "jolt," so how often a client jolts is a good indicator of how often a cleaner fish cheats. The predator clients don't jolt as much during cleaning as the harmless clients do.

What can the clients do to prevent their cleaner partner from cheating? First of all, as mentioned in Chapter 5, clients tend to go to cleaners that have a reputation for not cheating. Second, if they are fooled and go to a cleaner with a good reputation who then cheats, they can try to punish the cheater by attacking or trying to eat it. If the cheater is eaten, you might say that the punishment is effective. But even if the cheating cleaner is simply chased away by the client, this seems to affect the cleaner's future behavior. Clearly, when "thinking" about cheating, the cleaners take the possibility of punishment into consideration. If the punishment option is eliminated, cheating gets out of control. This was shown by an experiment in which clients were slightly anesthetized; with the client half-asleep, the cleaners cheated like crazy and mainly fed on their mucus and tissues instead of removing parasites.

Biological market theory predicts that nonpredatory clients have to accept more frequent cheating by the cleaner than predatory clients, and this turns out to be the case: as mentioned before, while in the cleaning station, the former jolted more frequently than the latter, and this was true for both resident and floater clients. Resident clients' punishment of cheating cleaners is about as effective as the switching strategy of choosy clients. In the language of game theory, the predator's option to kill the cleaner leads the cleaner to engage in an unconditional cooperative strategy with the predator. So, biological market effects influence partner choice, but partner control mechanisms are also important. Specifically, partner choice options determine which pairs form first. When it comes to the frequency of cheating by cleaner fish, however, partner choice options are overrun by client control mechanisms: predatory clients are far less often cheated than nonpredatory clients, irrespective of choice options.

As the cleaner-client fish market illustrates, although the Prisoner's Dilemma and biological market models seem to address different problems involved in cooperative interactions, in reality there are obvious connections between partner control and partner choice issues. Many recent studies conducted with human subjects have shown that when individuals are allowed to choose a partner, as opposed to being paired with a random individual, they tend to be more trusting and trustworthy, to be more cooperative in the two-player Prisoner's Dilemma game, and to contribute more to the production of public goods.[17] This happens when voluntary partner selection is permitted because people who have a tendency to cooperate choose each other and exclude defectors. However, if partner selection is competitive (in other words, market forces are operating), there is pressure on defectors to behave more cooperatively in order to be attractive in the market, to be chosen as partners, and therefore to be given an opportunity to play in the preferred circles where the benefits of cooperation can be obtained. This has led to the interesting idea that combining opportunities for partner choice with a competitive market can facilitate the emergence of prosocial behavior—that is, altruistic behavior that is costly to the individual but beneficial to the group. Before elaborating on this point, however, I want to illustrate another interesting human market—the book author–agent/publisher market.

THE BOOK AUTHOR–AGENT/PUBLISHER MARKET

Although publishing books is a cooperative enterprise unique to humans, it works according to the same laws of biological markets that operate in animal societies. In this market, there are two main classes of traders: people who write books (authors) and people who run publishing companies (publishers). The commodities offered by authors are their manuscripts, with all of their ideas, stories, facts, and illustrations. The commodities offered by publishers include the printing equipment necessary to turn a manuscript into many copies of a book, but also the resources to promote and distribute the book to various outlets. These two classes of traders must necessarily cooperate with one another to do business (although authors increasingly self-publish their books). In some cases, a third class of trader, the literary agent, acts as an intermediary between the author and the publisher. Agents represent authors and help them find publishers and negotiate a good deal with them. For the purposes of this discussion, agents and publishers serve a similar function, so I refer to them interchangeably.

Within each class of trader, there is great variation in the quality of the commodity being offered, and therefore in the value of the trader as a potential cooperation partner. Among authors, there are those who write unpublishable gibberish and those who write best sellers that sell millions of copies; similarly, there are more and less desirable publishers and effective and ineffective agents. In the United States alone, hundreds of thousands of people churn out manuscripts every year, but most are never published. Only a small minority of manuscripts are turned into books, and a tiny fraction of them become best sellers. Despite the low probability of writing a best seller, people continue writing manuscripts for the same reason they play the lottery: book publishing is a winner-take-all market, and the appeal of becoming an instant millionaire prompts people to write books despite the odds.

Not surprisingly, the market is skewed: authors greatly outnumber agents and publishers. In most cases, therefore, agents and publishers are the choosing classes and the authors are the chosen class. Authors compete with one another to find agents and publishers who are willing to trade with them, flooding their mailboxes with manuscripts and proposals that invariably promise to deliver a best seller. The vast majority of these works end up being rejected after a cursory examination. Agents and publishers, too, compete for access to a few best-selling authors who can make them a huge profit. Thus, there is competitive partner selection on both sides of the market, with a heavy skew against authors.

As in any other biological market, the quality of commodities, and therefore the value of the traders, is determined by supply and demand, which can fluctuate over time. What determines market value for an author? One would be tempted to say the quality of the product, but that's not always the case. For many reasons—some objective and understandable (the amount of advertisement and promotion of the book) and others arbitrary and uncontrollable (people's reading preferences and societal trends)—some pretty bad books end up becoming best sellers while some marvelous works are never published or simply ignored. To give a few examples, according to the website Just My Best, Robert M. Pirsig's novel *Zen and the Art of Motorcycle Maintenance* was initially rejected by 121 different publishers before becoming a huge best seller: more than 5 million copies were sold worldwide. When John Grisham's novel *A Time to Kill* was rejected by fifteen publishers and thirty agents, Grisham ended up publishing it himself. Other, even more famous self-published books that garnered multiple rejections include *Ulysses* by James Joyce and *Remembrance of Things Past* by Marcel Proust. According to Just My Best:

> Stephen King's first four novels were rejected. "This guy from Maine sent in this novel over the transom," said Bill Thompson, his former editor at Doubleday. Mr. Thompson, sensing something there, asked to see subsequent novels, but still rejected the next three. However, King withstood the rejection, and Mr. Thompson finally bought the fifth novel, despite his colleagues' lack of enthusiasm, for $2,500. It was called *Carrie*.[18]

As these examples illustrate, agents and publishers do not necessarily base their acceptance or rejection decisions on quality. They make decisions based on the probability that the book will sell, and book quality alone is not a good predictor of success. Two other predictors of success for a book, and of the market value of its author, are better: whether a previous book by the same author was a best seller, and whether the book is on a topic in which there is a lot of interest. Once readers develop an interest in an author or a topic, the thinking goes, they will buy any books from that author or on that topic, regardless of quality.

What determines quality for literary agents? It's mostly their previous success and reputation. A few agents are extremely successful and therefore have high value as trading partners. For example, in the field of scientific books written for general audiences, one particular literary agent is considered very successful and is highly sought after by authors. As the representative of many best-selling authors, he is able to put up for auction among publishers almost any book he manages, knowing that publishers will try to outbid each other to secure these books. As a result, his authors receive very profitable advance payments, and their books are more likely to be successful. The quality of a publisher can have an important impact on the success of a book, and a publisher that has previous success in publishing best sellers, name recognition, and the financial resources to pay large advances and promote books is a high-quality publisher.

Given the determinants of market value for an author, a first-time author who writes a book on a topic for which there is low demand is in bad shape. The main topic of my book *Macachiavellian Intelligence: How Rhesus Macaques and Humans Have Conquered the World*, the first scientific book for general audiences that I wrote, is the behavior of rhesus macaques—not a hot topic for the general public by any stretch of the imagination. Not surprisingly, I had trouble finding partners for cooperation among agents and publishers, so I traveled to publishing Bangkok: I moved into a different market where my endowments were more valuable—the academic book publishing market. University presses mainly publish scholarly books that cover narrow topics and are read by very few people. In addition, most university presses, lacking the financial resources, do little or no book promotion, and in the larger book publishing market this is the kiss of death. As a result, books published by university presses sell, on average, only a few hundred copies, and any book selling more than a thousand copies is considered successful and profitable. In the market of academic book publishing, a scientific book written for a general audience, which can easily sell more than a thousand copies, is considered a valuable commodity, and professors who write such books are highly sought-after trading partners. In this new market, I did pretty well: despite my old age, my baldness, and my belly—I am speaking figuratively, of course—I was able to find a young and attractive wife after all.

In the book publishing market, as in the primate grooming market, temporal fluctuations in supply and demand can dramatically change the value of certain commodities and of the traders who possess them. For example, twenty years ago books for general audiences about the relation between economics, psychology, and people's behavior were not in demand at all and were therefore mainly published by university presses. The amazing success of *Freakonomics* by Steven Levitt and Stephen Dubner, however, as well as books by Steven Pinker and Malcolm Gladwell, greatly increased the demand for this kind of book, and their market value increased dramatically. Agents and publishers became extremely interested all of a sudden in doing business with economics and psychology professors who could write books for general audiences. As the supply of these books increases and the demand decreases (owing to the fading of the novelty effect that was largely responsible for their success), more and more of these new books, some of which are superior in quality to their best-selling predecessors, will end up in the Bangkok market of academic publishing.

Despite the fact that the book publishing market is driven by financial profit and not by other goals such as survival or reproduction, it is effectively a biological market like the mating markets in humans and animals and mutualistic markets involving organisms of different species. In the book publishing market, traders with commodities that differ in value shop for cooperation partners through competitive partner choice mechanisms and according to the laws of supply and demand. This market may be about turning objects (books) into money (and/or power and fame), but just like the business of scientific research, peer review, and grants and publications, it involves people negotiating with other people and therefore follows the models of competitive and cooperative social behavior developed by evolutionary biologists and economists.

Now, let me return to the idea that the combination of opportunities for partner choice and a competitive market can facilitate the emergence of prosocial behavior. Evolutionary psychologist Yen-Sheng Chiang at the University of California–Irvine published an article in 2010 in which he reported the results of an interesting study showing that competitive partner selection facilitates the emergence of fairness when people play the Ultimatum Game, a two-player economic game similar to the Dictator Game described in Chapter 5.[19]

The first player (the "proposer") makes an offer to divide a certain amount of money to the second player (the "respondent"), who then decides whether or not to accept the proposal. This study compared people's behavior in two different situations: being able to choose their partner, and having their partner assigned randomly. Chiang wanted to know whether the offers made in the selected-partner situation were fairer than those in the situation in which partners were randomly assigned.

The study involved fifty-eight undergraduate students at a large public university in northwestern America who were recruited and paired up to play the Ultimatum Game with one hundred chips (converted into real money after the experiment) to divide up. They played the game through networked computers. In the first five rounds—the standard treatment—subjects played an anonymous Ultimatum Game with partners who were randomly assigned by the computer and about whom they received no information. Starting in the sixth round, players entered a new situation—the partner selection treatment—for fifteen rounds. In each of the fifteen rounds subjects were given the history of play for each player in the other role and then asked to rank the players they would prefer to play with for the current round of the game.

Not surprisingly, when given the opportunity to rank their preferences for other players, proposers ranked very highly responders who had shown high acceptance rates in the past and who had not recently rejected low offers, while responders preferred to be paired with proposers who had made high offers in recent history. In other words, both proposers and responders preferred to be paired with players who behaved altruistically (the Ultimatum Game is a zero-sum game—if one player gets more, the other player necessarily gets less), and they wanted to do it for selfish reasons, to maximize their profit. Since everybody liked the same individuals to play with, partner selection became a competitive process. Not everyone, however, could play with their preferred partner, and therefore both proposers and responders had to outbid competitors within their own class to be attractive in the market. In the partner selection treatment, proposers made, on average, fairer offers than in the standard treatment (46.28 versus 42.20 chips). So the experiment nicely illustrates how fairness can emerge out of selfishness when partner selection takes place in a competitive market.

Ronald Noë has extended this idea even further.[20] He argues that whenever single powerful individuals (for example, village chiefs, kings, warlords, or priests) or institutions (councils of elders or political parties) favor group members on the basis of their prosocial behavior—their tendency to sacrifice themselves for the sake of the group, that is, to be good "team players"—the evolution of prosocial and altruistic behavior in general is encouraged. During recent human evolution, the selection of altruistic individuals as team members may have taken place numerous times during the formation of hunting parties, raiding teams, military regiments, and the like. From an evolutionary standpoint, choosing an altruistic individual as a favorite partner for cooperation can be favored by natural selection only when it brings benefits to both the choosing individual and the chosen one. For example, the leader of the hunt should be able to obtain a larger amount of meat at the end of the day by choosing the right hunters for his party, and the chosen hunters should benefit more than those who are excluded. Similarly, when a team is formed with the goal of producing public goods, such as cleaning train stations or protecting the environment, the selection of members on the basis of characteristics that make them good team players (such as loyalty to the team, willingness to back up failing teammates, or fairness in sharing) can, in the long run, encourage the expression of prosocial behavior in all individuals.

Noë remarks that such traits are still highly relevant in modern societies: "Being a team player is of paramount importance in the workplace, according to both employers and employees. Being perceived as a team player is considered to be more important than doing a good job, being intelligent, being creative, making money for the organization, and having many other good qualities." Noë also suggests that the characteristics that make individuals good team players may be especially encouraged—and if they have a genetic basis, they could be favored by natural selection—in despotic societies ruled by a strong central power, such as a dictator, because in such societies individuals who respect authority and follow rules are rewarded, whereas individualistic personalities who challenge authority are penalized. However, Noë notes that team players can enjoy advantages in egalitarian societies as well, because in these societies they are often recruited and invited to join teams by other altruistic individuals. In all societies, humans are unique among animals in that our ability to report the performances of team members to the rest of the community increases the necessity to acquire a good reputation as a team player and reinforces partner preferences for altruistic individuals.

Since team-playing traits handicap the individual who possesses them but benefit the group to which he or she belongs, such traits, if genetically determined, could evolve by group selection, a controversial evolutionary process in which natural selection favors behavioral traits that are costly to individuals but beneficial to their groups. Noë's idea provides a mechanism for the evolution of prosocial behavior that does not require group selection. In his view, the costs of prosocial behavior to the individual (that is, the sacrifice he or she makes to benefit the group) are offset by the benefit of establishing a good reputation and of being chosen as a team member.

The only problem with the proposed role of partner choice in the evolution of human prosocial behavior is that this mechanism can also work in the opposite direction. Powerful individuals and institutions that are in competition, or at war, with other powerful individuals and institutions can select partners who make great fighters: for instance, selfish, ruthless assassins and mercenaries who don't hesitate to kill others for self-defense or personal profit. Since competition between groups probably played as important a role in human social evolution as cooperation within groups, partner choice for selfish, competitive, and aggressive individuals may have represented a powerful evolutionary force operating in opposition to the forces promoting the selection of altruistic and prosocial behavior.

Chapter 9

The Evolution of
Human Social Behavior

Evolutionary Baggage

The biographies of famous artists, musicians, scientists, philosophers, spiritual leaders, and other remarkable individuals provide unique insights into human nature. The accomplishments of these individuals influenced the lives and work of millions of other people. Just think about how many human lives have been touched by Mozart and Picasso, Einstein and Darwin, Plato and Aristotle, or Gandhi and Mother Teresa of Calcutta. Yet, if one examines the *social* lives of many of these intellectual and spiritual overachievers, they come across as far less virtuous and remarkable than their "professional" legacies would have us believe.

Spanish painter Pablo Picasso took advantage of his fame and artistic talent to sleep with almost every woman he met in his adult life (including a seventeen-year-old model he met when he was forty-five and with whom he later had a child). According to biographer Patrick O'Brian, Picasso married twice and had four children by three different women, and regardless of his marital situation, he always kept several mistresses in the background.[1] Picasso was a tremendously productive artist—his oeuvre comprises more than fifty thousand paintings, drawings, and sculptures—but work was clearly not the only thing he had on his mind. In that respect, he is in good company. Hundreds of thousands of men who have achieved fame through art,

229

music; science, or other intellectual activities also cheated on their wives and used their fame to maintain harems of women to satiate their voracious sexual appetites.

Just as some great minds are susceptible to the temptations of promiscuous sex, others are lured by the prospect of political power. To give one example close to my professional home, Austrian ethologist Konrad Lorenz, who won a Nobel Prize in 1973 for his research on animal behavior, was unpopular with many of his colleagues and for a long time could not find an academic job in his country, confirming the Latin proverb *Nemo propheta in patria*, loosely translated as "Nobody is appreciated in his homeland." Science historian Richard Burkhardt—author of the 2005 book *Patterns of Behavior: Konrad Lorenz, Niko Tinbergen, and the Founding of Ethology*—believes that Lorenz joined the Nazi party in 1938 so that he could secure an academic job in Germany, which he did: in 1940 the Nazi regime arranged a professorship for him at the University of Königsberg.[2] Lorenz was not the first intellectual to strike a deal with a dictatorial regime—or more generally, to align himself with political power to advance his own career. He followed an illustrious tradition that originated in the ancient world and became well established in Europe during the Renaissance: scientists, artists, and musicians seeking patronage from emperors, kings, and popes and often not only obtaining employment, support, and protection in the process but also amassing a great deal of political power themselves.

Finally, many spiritual and religious leaders who encouraged their followers to live a virtuous life simultaneously showed a keen interest in the material benefits of this world. Mother Teresa of Calcutta, an Albanian Roman Catholic nun who received the Nobel Peace Prize in 1979 and was declared a saint by Pope John Paul II in 2003, did not hesitate to support wealthy and corrupt individuals, including Haitian dictator Jean-Claude Duvalier and former financial executive and white-collar criminal Charles Keating, to gain millions of tax-free dollars from them. In his 1997 book *The Missionary Position: Mother Teresa in Theory and Practice*, author and columnist Christopher Hitchens argues that Mother Teresa was less interested in help-ing the poor than in stashing away vast amounts of cash with which to fuel the expansion of her fundamentalist Roman Catholic beliefs.[3]

Clearly, exceptional individuals—people with superior education, intelligence, artistic talents, or religious and moral principles—share many traits with the rest of the human race: social and political am-bitions, greed for money, rivalries with contemporaries, unrestrained sexual appetites, and marital problems. In many cases, there seems to be a sharp disconnect between the intellectual or spiritual achieve-ments of these famous individuals and the content and quality of their social lives. Why this disconnect?

The answer, I think, is that human social behavior comes with a heavy load of evolutionary baggage—we all have strong biological predispositions to behave in certain ways and to pursue similar goals in our personal lives. In the end, we all want the same things: money, power, fame, sex, love, and children. In contrast, our intellectual po-tential is almost infinite and can be realized in a thousand different ways—or not realized at all. What we do with our intellect has little to do with our evolutionary past and our biological predispositions and a lot more to do with our environment, our education, and the opportunities that our lives present to us. In theory, given the right environment, anyone can become an accomplished painter, musician, philosopher, or theoretical physicist. Some exceptional individuals advance and specialize so much in these domains that laypeople can't even begin to understand their accomplishments. How many of us can confidently state that we fully appreciate the extent of Albert Einstein's contributions to physics or those of Ludwig Wittgenstein to philosophy? However, all it takes to fully understand the social lives of many Nobel laureates is some knowledge of primate social behavior. The social behavior of intellectual and spiritual leaders (as well as that of kings and emperors, politicians and army generals, and rock-and-roll, movie, and sports celebrities) is generally similar not only to the behavior of non-achievers but also to that of monkeys and apes and other animals as well.

Anthropocentrism and Free Will

A friend of mine who read my book *Macachiavellian Intelligence: How Rhesus Macaques and Humans Have Conquered the World* and didn't know much about the social behavior of rhesus macaques beforehand had the following reaction: "Wow, these monkeys really behave like people. They *are* people!" To which I replied: "No, it's people who really behave like other primates. People *are* primates."

Our species, *Homo sapiens*, belongs to a group of mammals called primates and, more specifically, to a subgroup of primates called great apes. There used to be many species of great apes on our planet, but many of them have gone extinct. The remaining great apes are the chimpanzees, bonobos, gorillas, and orangutans. Our closest relatives are chimpanzees and bonobos, which share with us about 98 percent of their genetic material. Closely related to us and the other great apes are the lesser apes—gibbons and siamangs—and Old World monkeys such as macaques and baboons, which share with us about 95 percent of their genetic material. Studies of fossils and comparisons of DNA between different species suggest that our hominid ancestors split from the ancestors of the other great apes between 5 and 6 million years ago, from those of gibbons and siamangs about 10 million years ago, and from those of macaques and baboons about 25 million years ago.

The taxonomic classification of human beings was established on the basis of anatomical similarities long before genetic data were available and long before Darwin published his evolutionary explanations for these similarities. The taxonomic classification of human beings is not particularly controversial. Even creationists who believe that evolution is just a theory (it's not: evolution is a fact, as nicely illustrated by evolutionary biologist Richard Dawkins in his 2009 book *The Greatest Show on Earth: The Evidence for Evolution*) don't seem to challenge our taxonomic status as a primate species.[4] Who cares about taxonomy anyway? It's just a bunch of labels, isn't it?

But we do care about other aspects of our "humanness." My friend's surprised reaction upon discovering the similarities between human and rhesus macaque behavior is representative of the way a lot of people think about themselves and their behavior. First, there is the "they are like us" versus "we are like them" issue. This has to do with anthropocentrism—the idea that humans are at the center of the universe and everything else revolves around them. Telling some people that there are primates out there, such as rhesus macaques and chimpanzees, that are very similar to us in behavioral terms is like telling Ptolemy that two new planets have been discovered in the Earth's orbit. Our solar system becomes a little larger, but we are still there, right in the center of it. We are the sun, and all the others are planets, regardless of how many of them there are out there.

Anthropocentrism is stronger for some human traits than for others. People's faces and bodies resemble the faces and bodies of other animals, but when it comes to these similarities, no one cares about the distinction between stars and planets. The Walt Disney Corporation and toy manufacturers around the world have made billions of dollars exploiting animal-human similarities for their cartoons and stuffed animals. Not everyone may know that our bones, muscles, skin, hearts, lungs, and stomachs and intestines work exactly like those of other animals, but I am pretty sure that many people given this news would shrug their shoulders and say, "Okay, so what?" When we discover similarities in behavior, however, our anthropocentrism kicks in. We are not like them. They are like us. Maybe.

Similarities in behavior between humans and other animals are a source of endless controversy. People generally have a different view of their behavior than they do of their faces and bodies—as if bodies are biological but behavior is something special, something nonbiological. I suspect that one of the factors at play is the question of free will. We were born with our particular faces and bodies, and until the advent of plastic surgery there was nothing we could do about it. Now, as long as we can afford it, we can have almost any face or body we want. We don't need any professional nip and tuck for our behavior, however, because we perform our own cosmetic surgeries every day. We wake up in the morning and make plans for the day. Then we change our minds and make different plans, sometimes more than once. Everything we do we think about first, and then we do it. Thinking is the cause and behavior is the effect—or so we believe. We make hundreds of conscious decisions about our behavior during the course of a day. How can this be affected by millions of years of evolution? How is it possible that the product of free will ends up resembling what monkeys and apes in the jungle have been doing for millions of years?

Free will and anthropocentrism go hand in hand. The seventeenth-century French philosopher Descartes, who came up with the brilliant phrase *Cogito, ergo sum* ("I think, therefore I am"), believed that humans are unique in possessing free will and that all other animals act like robots. Accordingly, Descartes placed humans right at the center of the universe. Many embraced Descartes's view, including the psychologist William James, who wrote in 1890 that the whole "sting and excitement" of life comes from "our sense that in it things are really being decided from one moment to another, and that it is not the dull rattling off of a chain that was forged innumerable ages ago."

Well, for over two centuries people have had to disabuse themselves of the notion that humans are fundamentally different from other animals. Beliefs about the uniqueness of human behavior might well be the last bastion of our superiority complex, but even this redoubt may be crumbling. As for free will, experiments in psychology and neuroscience suggest that—to paraphrase *The Princess Bride*—"I don't think it is what you think it is."

In a 2007 *New York Times* article entitled "Free Will: Now You Have It, Now You Don't," science columnist Dennis Overbye (currently the deputy science editor of the *Times*) reported on interviews with two scientists—Benjamin Libet, a former physiologist at the University of California–San Francisco (he died in 2007), and Daniel Wegner, a psychologist at Harvard University—who have done research on the issue of free will.[5] In the 1980s, Libet conducted experiments in which he asked volunteers to choose a random movement with their hand, such as pressing a button or flicking a finger, while he recorded electrical activity in their brains through an electroencephalogram. Libet asked the subjects to watch the second hand of a clock and report its position at exactly the moment they felt they had the conscious will to move. The experiments showed that a spike of electricity occurred in the neurons in the brain that control hand movement approximately half a second before the subjects consciously felt that they had decided to move their hand. In other words, it appeared that the brain unconsciously controls behavior before a conscious decision is made. The perception of an action makes an individual conscious of it, and this after-the-fact consciousness generates the illusion that our behavior is the result of a decision and not the other way around. Libet's experiments suggested that free will is an illusion—a trick played on us by our own minds. His results have been replicated by other neuroscientists (although, as usual, some skeptics have criticized them), while experiments by Daniel Wegner, summarized in his 2002 book *The Illusion of Conscious Will*, have shown that people can be easily fooled into believing that they cause and control their own actions when in fact they don't.[6]

In the interviews for the *New York Times* article, Libet said that his results left room for a limited version of free will in the form of veto power over what we sense ourselves doing—we can choose to inhibit our behavior once we become conscious of it—while Wegner commented on the potential consequences of exposing free will as an illusion. Some people worry, he said, that the death of free will could wreak havoc on our sense of moral and legal responsibility: people might feel that they are no longer responsible for their actions. Wegner believed, however, that in reality exposing free will as an illusion probably would have little effect on people's lives or on their feelings of self-worth. Most people would remain in denial. "It's an illusion, but it's a very persistent illusion; it keeps coming back," he said, comparing free will to a magician's trick that has been seen again and again. "Even though you know it's a trick, you get fooled every time. The feelings just don't go away."

The Algorithms That Crowd Our Minds

Many people reject the idea that human social behavior is hardwired in our brains and evolved by natural selection earlier in human evolutionary history or in the history of our animal ancestors. Some object on religious grounds, such as the creationists, many of whom believe that human beings were created by God, as described in the Book of Genesis. There are also intellectuals—some cultural anthropologists and psychologists—who dismiss biology and evolution as irrelevant to our understanding of human behavior; they argue that cultural influences on behavior override biological ones.

Surprisingly, the idea that behavior evolves is also unpopular among evolutionary biologists who study genes, cells, or tiny insects called fruit flies. They keep hundreds of fruit flies in glass jars in their laboratories, induce some changes in their environment—such as turning the lights on and off for variable periods of time—and then look at how the fruit flies' genes are affected by this as the bugs reproduce over many generations. These evolutionary biologists study evolution in their laboratory but not in the real world or with animals other than fruit flies. The scientific journal *Evolution*, for example, publishes many articles involving experimental studies of evolution conducted with fruit flies. I once submitted a manuscript to this journal reporting an evolutionary analysis of primate behavior, and the manuscript was immediately rejected on the ground that "primates are non-standard organisms for the study of evolution." "I don't think Charles Darwin would agree with you," I replied to the journal's editor.

For some evolutionary biologists, evolution stays confined within the walls of their laboratory. For others, it stops at their doorstep: it can happen in the jungle, but they don't want it in their house. They don't want to hear about how evolution affects their own behavior. They are like those Catholics who go to Mass every Sunday but, forgetting all about religion once they're outside the church, spend the rest of the week going about their business as usual. Surprisingly, among the skeptics are also some evolutionary psychologists who believe that natural selection shaped the human mind, yet maintain that what we do with our behavior has little to do with evolution. I for one have a hard time accepting the notion that natural selection has left its mark on human mental processes but not on contemporary human behavior. The latter happens to be the premise upon which this whole book is based! Before I address the position taken by evolutionary psychologists about the evolution of mind versus behavior, however, let me talk a bit more about evolutionary psychologists in general, and about how the human mind works.[7]

Evolutionary psychologists don't believe that the human mind is a *tabula rasa*—an empty container we fill with all kinds of information that we acquire from the environment with our amazing learning skills. Rather, they believe that the mind has some biological predispositions to produce specific emotions over others in response to particular situations, to learn certain things better than others, to solve problems in a certain way, and even to make certain mistakes in the perception and processing of information from the surrounding environment. For example, children are biologically predisposed to learn languages until they reach puberty; after that, changes in their brain make it very difficult to learn a new language—as people who have learned a second language later in life know all too well. Studies by social psychologists have shown that people generally have a better view of themselves than others have of them. We see ourselves as being nicer, more intelligent, and more successful than others do. The human mind is the device that drives our bodies in the race for survival and reproduction, so it makes sense that it gives us the impression that we are at the center of the universe—anthropocentrism is a psychological adaptation—and that it has many mechanisms to protect us from the challenges, not only physical but also psychological, that come from the outside world. Sigmund Freud, a man with a brilliant intuition who had the bad luck of being ahead of his time, discovered many of these protective mechanisms, such as the repression of bad memories.

Called *algorithms* by evolutionary psychologists, the human mind's predispositions are akin to computer programs in that they were designed to solve particular problems or tasks. They have a significant genetic basis and evolved by natural selection, so that individuals with the genes for particular algorithms were more successful in survival and reproduction than the individuals without these genes. Algorithms represent solutions to the recurring problems that early humans and their ancestors were constantly confronted with in their environment, including: how to navigate and orient oneself; how to find food and discriminate between edible/nutritional substances and toxic/non-nutritional ones; how to detect predators and escape predator attacks (as well as avoiding dangerous animals such as venomous snakes and spiders); how to avoid a potentially deadly aggression from a member of another group or harmful and coercive behavior from a member of one's own group; how to learn to communicate with members of one's own community; how to discriminate family members from nonkin; how to make friends and practice important social skills with them; how to establish cooperative relationships with others based on reciprocation and how to identify and punish cheaters; how to make effective political alliances that allow one to outcompete other individuals and gain status; how to find and choose appropriate and willing mates for short-term sexual relationships; how to establish a long-term relationship with an opposite-sex individual that will lead to successful reproduction and child-rearing; how to support and shape children's development so that they will become successful adults; and how to obtain assistance and support from caregivers early or late in life, periods when individuals cannot make it on their own.

Many, if not all, of these problems have been recurring throughout the lives of human beings, and finding an appropriate solution for them can mean the difference between life and death, or between having many grandchildren and having none. It is difficult to imagine that all human beings must figure out solutions to these problems on their own, learning everything from scratch over the course of their lifetime—often with few or no opportunities to gain experience beforehand—and picking the most appropriate solution to the problem among the myriad possible different options. No, natural selection gives us a hand by suggesting successful solutions found by our ancestors. In some cases, the correct behavioral response to the problem is entirely hardwired in our brain and automatically triggered whenever we confront the problem; in others, we need to figure out the solution on our own, but our predispositions give us a big push in the right direction. Clearly, when we say that a trait—bones, body organs, or behavior—is the product of natural selection, we imply that the trait is genetically controlled. This doesn't mean, however, that there are single genes for specific algorithms—psychological and behavioral algorithms may be influenced by many genes interacting in complex ways—or that the environment is not important. In reality, traits are always the result of interactions between genes and the environment. One cannot work without the other.

The kind of algorithms we possess in our minds depends on the kind of problems that recurred in the environment of our ancestors. If humans had been small fish, we would be born with all kinds of predispositions for swimming, navigation, and orientation in water; avoidance of sharks and other predators; coordination of movement with other fish; and so on. But humans are highly social primates, so we must deal with the ecological problems typical of other primates—spatial orientation in forests or vast open spaces, finding food appropriate to our digestive physiology, avoiding dangerous animals—plus the host of problems that arise from living long lives in complex and highly competitive societies. No wonder, then, if many of our mental and behavioral algorithms have something to do with solving social problems.

But what exactly are these algorithms, and what do they do for us? Structurally, they are complex neural circuits, some of which are located in particular brain regions and others of which are diffused throughout the entire brain in a thick web of intricate neural connections. In humans, research with a technique called functional brain imaging—which allows researchers to obtain visual images of the brain areas that get activated when we experience particular thoughts or solve particular behavioral tasks—is beginning to tell us what these algorithms are and where they are, but this is an extremely difficult enterprise, and we still know very little about it. We know a lot more about the function of algorithms: what they do for us and how they operate.

Some of these predispositions are simple preferences for certain visual, auditory, gustatory, or olfactory stimuli over others. Human babies are predisposed to show interest in faces, and especially in the eyes of faces. Sexually mature heterosexual males are predisposed to be visually attracted to female faces with infantile characteristics (which suggest a young age) and to female bodies with a thin waist and wide hips (which suggest high fertility). Females of all ages are predisposed to be visually attracted to baby faces. Infants and children are predisposed to be attracted to the sounds of baby talk, or moth-erese. We are all predisposed to like the taste of sugar. Other algorithms take the form of predispositions to experience particular emotions in response to particular situations, to make particular decisions when given a particular set of options, and more generally to act in particular ways with particular individuals and situations. Emotions represent very powerful biological predispositions to help us deal with environmental problems and are involved in virtually all of the social situations I discussed earlier in this book.

Emotions: Activators and Coordinators of Programs

When ecological and social problems arise, we have some behavioral algorithms at our disposal to help us solve them. But natural selection doesn't just provide us with the solution to the problem; it also makes sure that we actually use it, and sometimes that we use it quickly. Behavior doesn't always happen directly in response to the external environment. Rather, behavior is activated by a trigger inside our body. The environment pulls our internal trigger, and the trigger activates behavior. This trigger is called *motivation*. Motivation, or "wanting to do something," doesn't necessarily imply consciousness and free will. We can be motivated to do something and be totally unaware of it. In many cases, the substrate for motivation is a physiological reaction that begins in the body, where an event from the environment is recorded, and then moves into the brain. To give an example, if you stick your index finger in a flame, the flame will burn your skin and hurt the nerve cells in your fingertip. Pain, a physiological reaction that results from a connection between body and brain, is the trigger, the motivation that makes you withdraw your finger quickly.

Emotions are similar to pain, but more complex. As mentioned in Chapter 6, one function of emotions is to energize motivation, but they do a lot more than that. According to evolutionary psychologists John Tooby and Leda Cosmides, emotions are "computer programs" designed by natural selection not only to motivate behavioral responses but also to coordinate and organize other algorithms or subprograms. It is often the case that some of these algorithms need to be activated and others need to be deactivated in order for the appropriate behavioral response to be expressed, because at any point in time our body is simultaneously confronted with multiple different problems that require an adaptive solution.[8]

Imagine that you are walking alone on a street late at night. You haven't had dinner yet, and the sight of people eating burgers at a McDonald's and the smell of French fries tell you that you have a problem. The motivational trigger—hunger—is activated, and you want to find food. But you also haven't had sex in months, and the sight of an attractive, half-naked model on a giant Victoria's Secret lingerie billboard ad tells you that you have a problem. The trigger—lust—is activated, and you want to find a mate. But you also haven't slept in three days, and walking nonstop for five hours has reinforced the notion that you have a problem. The trigger—fatigue—is activated, and you want to crash on a bed. But all of a sudden you feel the barrel of a gun pressed to the back of your head and a voice saying: "Give me your wallet or I will kill you." Your life is in danger, so you definitely have a problem. The trigger—fear—is activated, and it makes you want to run away as fast as you can. But what about that Big Mac and fries, the lingerie-clad Victoria's Secret model, and the comfortable bed that you wanted so badly a second ago? Luckily for you, your fear of being killed suppresses your hunger, lust, and fatigue so that the behavioral algorithms activated by these other triggers are not in competition with the one you now need the most to save your life. Eating a burger, having sex, or falling asleep is the last thing you want to do when you have a gun pointed at your head. Instead, other cognitive processes are activated: you are very alert and start processing other information from the environment and from memory. From the intensity with which the gun is pressed to your head, you get a sense of how potentially ready the robber is to kill you. From his voice, you gauge his size, his anger or fear, and his overall dangerousness. You estimate the distance from the nearest street corner and the time it would take you to get there if you started running. You remember that there is a garage where you could hide just around that corner. Finally, you scan the airwaves with your ears hoping to hear sirens in the background, which may suggest a police car is nearby. If you knew that the robber is your best friend playing a practical joke on you with a toy gun, none of this would happen. You would just laugh and keep thinking about the burger and all the rest. Instead, because the danger is real, your fear has all of these effects on your sensory processing, your memory and cognitive evaluations, and your shifting goals and motivations.

In the words of Tooby and Cosmides, the human mind is "crowded with functionally specialized programs." In addition to behavioral programs, there are cognitive and physiological programs that regulate the functions of our minds and bodies. According to Tooby and Cosmides, these programs regulate, among other things,

> perception and attention; inference; learning; memory; goal choice; motivational priorities; categorization and conceptual frameworks; physiological reactions (such as heart rate, endocrine, immune, and reproductive function); reflexes; behavioral decision rules; motor systems; communication processes; energy level and effort allocation; affective coloration of events and stimuli; recalibration of probability estimates; situation assessments; and values and regulatory variables (such as self-esteem, estimation of relative formidability; relative value of alternative goal states; efficacy discount rate).

In the academic jargon of Tooby and Cosmides, "Programs that are individually designed to solve specific adaptive problems could, if simultaneously activated, deliver outputs that conflict with one another, interfering with or nullifying each other's functional products. To avoid such consequences, the mind must be equipped with superordinate programs that override and deactivate some programs when others are activated." Such "superordinate" programs are our emotions. Their function is to direct and coordinate the activities and interactions of all the other behavioral, physiological, and cognitive subprograms.

This might all sound a little complicated. Does it really work this way? Do emotions really guide us to do the right thing in every circumstance? Suppose that while you are being robbed at gunpoint your fear activates the fleeing response but, as you try to escape, the robber shoots and kills you. Was the wrong emotion or the wrong response activated? Did natural selection screw up? No, it would be unfair to expect natural selection to get it right every time. Natural selection that operated thousands or millions of years ago could not predict the contingencies and outcome of every individual situation in the future. Each emotional program was selected to activate subprograms that, *when averaged over individuals and generations*, would have led to the best course of action. Every situation is unique, however, and there is some uncertainty linked to the contingencies of the environment. In the words of Tooby and Cosmides, "An emotion is a bet placed under conditions of uncertainty. Running away in terror, vomiting in disgust, or attacking in rage are bets that are placed because these responses had the highest *average* payoffs for our ancestors, given the eliciting conditions."

In some cases, of course, being held at gunpoint results in being shot, not because the right behavioral response doesn't work but because the wrong emotion was elicited, or because no emotion was elicited when one should have been. We routinely find ourselves in evolutionarily novel situations for which natural selection has not prepared us to respond appropriately, and when we do, trouble or maladaptive behavior may arise. Evolutionary psychologists call this situation a "mismatch" between our evolved emotional responses and the novel environment in which they happen to be elicited. A classic example is crossing the street and not being afraid of being hit by a car. Automobiles kill pedestrians all the time, but because cars are a recent phenomenon, we don't respond to risky situations involving them with the appropriate fear and anxiety. In contrast, we do have the propensity to be afraid of situations and stimuli that were associated with danger during much of our evolutionary past: darkness, heights, large predators, snakes, and spiders.

Even in novel situations, however, there may be cues that remind us of environmental problems that recurred billions or trillions of times during our evolutionary history. Some of our ancestors dealt with these problems well, while others didn't; thus, some survived and reproduced, and others didn't. We are the descendants of the successful problem-solvers. Thanks to natural selection, we have an innate knowledge of these recurring problems and the situations in which they arise; we have responses to deal with these problems and mental programs that help activate and guide these responses. Each emotion evolved to deal with a particular evolutionarily recurrent situation type. When familiar cues of danger are perceived—and this can happen unconsciously—a particular emotional trigger is immediately activated.

Let's go back to the elevator ride with a stranger in Chapter 1. This is an evolutionarily novel situation, like crossing the street at a busy intersection. Natural selection didn't prepare us to deal with elevator rides, and it certainly didn't give us an elevator-specific emotion or elevator-specific behavioral algorithms. Nevertheless, the elevator situation, though novel itself, is full of cues of danger that are not novel at all. First, you find yourself in close proximity to a stranger, a position that is a strong cue for risk of aggression. Second, you find yourself locked in a small space, so if the stranger attacks you, there is no opportunity to escape or to call for help from family members or allies. If the stranger attacks and you fight back in that small space, it is almost certain that you are going to get hurt. And finally, in addition to all this, if the stranger makes eye contact with you and stares you down, well, that's as strong a cue as you'll ever get.

This combination of cues has probably recurred millions of times during our evolutionary history, and natural selection has prepared us to deal with the situation and given us an emotion that will motivate and trigger appropriate behavioral responses. The emotion, however, is not fear. When you are being robbed at gunpoint, the threat is clear, the danger is high, and you could die within seconds. There is no uncertainty as to the intentions of the robber, and given the seriousness of the situation, fear is the appropriate emotion. Fear is reserved for emergency situations. In the elevator, however, there is a lot of uncertainty, both about the stranger's intention—is he indifferent, hostile, or friendly?—and about the possible courses of action: indifference, appeasement, or a preemptive threat. The appropriate emotion here is anxiety. A moderate level of anxiety will inhibit your motor behavior and make you avoid direct eye contact with the stranger.

As with fear, anxiety activates some motivational, cognitive, and behavioral processes and deactivates others. While avoiding eye contact, you may peek at the stranger standing next to you and unconsciously evaluate his formidability as a potential opponent relative to yours. Is he big and strong? Does he look like someone who could lose his temper quickly? Does he act with self-confidence, signaling that he is a high-status individual? How much social anxiety "leaks" from the stranger's behavior? As you make these assessments, you may unconsciously recall relevant memories about similar encounters with strangers with a similar profile, make comparisons, and estimate the outcome probabilities. High social anxiety may prompt you to take preemptive measures to minimize the risk of aggression, such as sending appeasing signals—smiling and making small talk.

Similar social anxiety and the accompanying cognitive and behavioral processes may also occur in a low-status individual who encounters a familiar higher-status person—for example, when you meet your company's boss in his office. In this case, your anxiety is based more directly on the awareness of the status difference, memories of past interactions, and their implications for the present and the future. The boss got mad at you and was verbally abusive in the past. Will he act similarly this time or not? You can't afford to lose your temper because the consequences could be devastating for your career. As with the stranger in the elevator, you unconsciously assess your boss's appearance and nonverbal behavior—for example, the way he looks at you or the tone of his voice—looking for potential cues of hostility. You process past and present information and make predictions about the future. Your anxiety doesn't just suppress behavior that might be misinterpreted as hostile, such as boldly looking your boss in the eyes. You also express explicitly submissive behavior such as avoiding his gaze, keeping your head down, smiling a lot, and speaking in a soft tone of voice.

The emotional flip side of fear and anxiety in socially dangerous or competitive situations is anger. Anger can motivate competitiveness and assertiveness in confrontations with individuals with whom you don't have a well-established dominance relationship. This could be anyone from an individual well known to you, such as your partner or spouse, to a stranger you are meeting for the first time or an out-group member such as a fan of a different football team or a member of a different political party. As discussed in Chapter 2, when a dominance relationship within a couple is not well established, confrontations are likely to be frequent and accompanied by outbursts of anger. If you get into a major car accident and the other driver starts screaming at you and accusing you of being at fault, you'd better yell back and hold your ground because the outcome of this confrontation may affect subsequent negotiations about assumption of responsibility and costs. Finally, confrontations with out-group individuals—whether they occur at sports events or at political rallies—are often accompanied by anger and animosity toward the members of the other group. In some types of confrontations with out-group members, fear is also elicited. Fear makes us infer danger and aggressiveness in others. Along with anger, fear can contribute to motivating competitive and aggressive responses toward out-group members. Generally, anger motivates assertive and threatening behavior, and sometimes acting self-confident or threatening is all that's needed to settle a contest to your advantage. An angry dog will growl and bark loudly to another one he meets in the park; if this display is effective, he becomes dominant over the other without any need for physical aggression. If a fight arises, between dogs or people, anger may provide the motivational fuel for fighting hard and long, or as long as it takes to win the battle. As with fear and social anxiety, when anger is triggered, some perceptual, cognitive, motivational, and behavioral subprograms are activated while others are deactivated.

Just as negative emotions such as fear, anxiety, and anger serve to protect us from danger or to increase our competitiveness, other emotions ensure that we are sexually attracted to another individual with whom we could potentially reproduce and that we form a long-term pair-bond that will allow for jointly raising children. Sexual and romantic feelings are elicited by cues of individuals and situations that, as with negative emotions, have recurred millions of times in our evolutionary history. We also have emotions to protect the exclusivity of our pair-bonds and prevent infidelity, such as sexual jealousy. Given the importance of parental investment for social success in complex human societies, parental love for children is an emotion that continues to be active and strong even when children become adults and is eventually extended to include grandchildren as well.

Finally, emotions play a role in establishing and coordinating the cooperative relationships with other individuals that are so important for survival and success in human societies. Natural selection has favored emotional processes that motivate and enhance an individual's ability to engage in, and profit from, cooperative enterprises. As nicely discussed by evolutionary psychologists Daniel Fessler and Kevin Haley in a 2003 essay, different emotions (and other psychological dispositions that may or may not fit the definition of emotions) affect cooperative behavior in different ways. Emotions such as trust, distrust, envy, and guilt play an important role in the implementation of cooperative strategies.[9] One needs to trust other individuals at some point in order to cooperate with them. Cosmides and Tooby have also argued that our mind contains a "cheater detection" algorithm that evolved specifically in the context of cooperation.[10] Anxiety or fear of being cheated triggers this cognitive subprogram that monitors the partner's behavior. Anger may also drive individuals to react strongly when they feel they have been cheated. Kindness and generosity play an important role in cooperative interactions as well. People respond with gratitude to spontaneous acts of generosity and feel compelled to reciprocate in kind. This might explain why subjects in behavioral economics games often violate the assumptions of traditional rational models by demonstrating a willingness to incur monetary costs in order to reward partners for perceived cooperative or altruistic behavior. Envy can play a role in market negotiations such as those discussed in Chapter 8: when actors identify a sizable disparity between parties in the possession of, or access to, valued goods or opportunities, those having less wish to obtain more. In addition to the desire to obtain what others possess, envy includes a measure of hostility toward the more fortunate party. Clearly, the emotions that operate in the context of cooperative interactions also operate in other social and nonsocial domains. When the same emotions operate in different domains, their effects are moderated by the characteristics of the context in which they occur.

Cognitive and Behavioral Algorithms

Until recently, many economists considered financial and other decisions to be the result of rational cognitive processes, and particularly of the rational evaluation of the costs and benefits of different options. Emotions, however, can change the subjective importance of costs and benefits, sometimes leading people to behave "irrationally." In some cases, emotions lead to irrational decisions because of a mismatch between emotion and situation. (The more a contemporary setting deviates from the ancestral environment, the less likely it is that such actions will be rational from an evolutionary perspective.) In other cases, people make seemingly irrational decisions not because their emotions interfere with rational decision-making processes but because emotions activate cognitive subprograms that lead to behavioral decisions different from those predicted by rational models.

Psychologists and economists used to believe that when people are presented with alternatives and must make a decision, they take all variables into account and behave in a way that maximizes their personal interests. It turns out, however, that in many situations people don't make decisions based on consideration and rational evaluation of all the information at hand. Rather, they use simple rules of thumb to make quick decisions in response to certain cues. Research by German cognitive psychologist Gerd Gigerenzer and others has shown that people possess "fast and frugal" algorithms, or heuristics, to make decisions in circumstances where economic theory has predicted sophisticated rational decisions. It turns out that the decisions made with fast and frugal algorithms are often more effective in dealing with certain situations and result in better outcomes than the decisions made through rational cognitive processes. In their 1999 book *Simple Heuristics That Make Us Smart*, Gerd Gigerenzer and Peter Todd give many examples of situations in which we use fast and frugal heuristics, from how we buy stocks to how we choose mates or divide resources among our children.[11] Incidentally, animals use heuristics too. For example, studies of animals and humans have shown that when individuals have to pick a mate among many possible partners, instead of using all the information at their disposal to evaluate the quality of each individual, they simply copy the choice of the majority of other individuals. Cognitive algorithms are probably the product of natural selection, which has prepared us to make quick and effective decisions in response to problems that have recurred many times in our evolutionary history.

Evolutionary psychologists are a lot more comfortable studying the emotional or cognitive programs of the human mind—how people feel and think—than the behavioral algorithms themselves—how people actually behave. They believe that natural selection shapes the mind's preferences, biases, and predispositions to respond to certain stimuli or cues, and that it's best to study emotional and cognitive algorithms in controlled laboratory conditions with a homogenous population of subjects, such as college students. In the lab, they typically look at how college students respond to visual stimuli, such as photos of people's faces or bodies, or how they solve simple cognitive tasks with paper and pen, or how they play computerized economic games.

Evolutionary psychologists rarely go out in the real world and observe how regular folks behave in their everyday lives. They believe that our behavior is too influenced by the surrounding environment, which is very different from the environment in which our minds evolved, for evolutionary psychology to make any sense of it. In short, they believe that it's difficult, if not impossible, to recognize and document the action of natural selection on the behavior of human beings living in modern industrialized societies. They may concede that studying primitive people such as the Yanomamo Indians of the Amazon or the !Kung of the Kalahari Desert has something interesting to tell us about human nature, but they are content to leave it at that.

Obviously I disagree. Despite the fact that we ride in elevators and communicate via email, the social problems and dilemmas we must confront in our everyday lives share plenty of similarities with those encountered during much of our evolutionary history. The cues we read and recognize in our everyday social situations don't simply trigger adaptive emotional and cognitive processes; they activate adaptive behavioral algorithms as well. The way we behave toward strangers in an elevator or during a meeting with our boss is the result of behavioral algorithms that are activated in situations where there is a high risk of aggression or we are confronting a high-status individual who has a great deal of influence over our life. The fact that human beings around the world act the same way in similar social situations suggests that a large part of our social behavior is genetically controlled. Sure, behavior is variable and can be influenced by the environment, but this variability is not infinite or arbitrary or unpredictable or maladaptive. We may not be able to predict the behavior of every individual in a particular situation, but we can predict what individuals will do in that situation on average.

When I was young, and long before I decided to specialize in study-ing monkey and human behavior, I spent a lot of time observing my pets, which happened to all be cats. Cats are interesting and behav-iorally complex creatures. If you don't believe me, watch a mother cat trying to teach her kittens how to hunt by bringing back prey to her nest and letting the kittens play with it. After observing not only my own cats but also those that populate the streets, squares, and an-cient ruins in Rome, I was struck by how homogeneous and stereo-typed cat behavior is. Sure, individual cats have different personalities. But there is huge uniformity in their behavior as well: there is defi-nitely a "cat nature," different from that of dogs and other animals. Apples come in all kinds of varieties too, differing in size, color, tex-ture, and taste. Yet, as a whole, apples are different from oranges. Cats are cats, apples are apples, and humans are humans. The differ-ences in behavior between animal species are the result of genetic, not environmental, differences. Domestic cats live in the same en-vironment as domestic dogs, and yet they still behave like cats and not dogs.

For pet lovers who remain unconvinced by my observations, con-sider the differences in behavior among different breeds of dogs. Clearly, there are strong differences among breeds: some dogs can easily be trained to retrieve or to help a shepherd keep his flock of sheep together. There are also differences in how aggressive or friendly different breeds of dogs are toward other dogs and toward people, how docile and cuddly they are, and how excitable or laid back. These differences are largely genetic and the result of selective breeding. To produce highly aggressive Dobermans, a dog breeder picks the most aggressive males and females he can find and allows them to breed, while not breeding the docile Dobermans. After generations of this selective breeding, Dobermans, on average, tend to be aggres-sive. Darwin used this example of behavior-based selective breeding of domestic animals extensively in *On the Origin of Species* to prove the point that behavior is heritable and that natural selection favors certain traits over others, the way an animal breeder does. Laboratory mice of different genetic strains also exhibit strong and consistent differences in emotionality and social behavior as a result of systematic selective breeding programs by researchers. Finally, thousands of recent studies in humans and all kinds of animals have shown that there is a correspondence between certain behavioral traits exhibited by individuals and the genes they possess. These traits include complex social behavior. Yes, of course, human behavior is still variable and influenced by early experience and environment. But exactly how variable or homogeneous human behavior is depends on one's perspective. To an anthropologist from Mars, all humans appear to act pretty much the same toward one another, and not at all like the way the inhabitants of Jupiter treat each other.

The Adaptiveness of Human Social Behavior and Convergent Evolution with Other Animals

Social behavior is, in part, genetically controlled and evolves by natural selection. Even though many of us live in technological and industrialized societies, much of our social behavior is still adaptive and solves the same problems it evolved to solve millions of years ago. For a highly social and competitive species such as *Homo sapiens*, the main source of problems—the main challenge to survival and reproduction—is not predators or lack of food or inclement weather, but other people. In response to the pressures to cooperate and compete, we behave nepotistically toward our kin at the expense of nonkin. We fight for dominance with enemies, friends, and family members, and depending on our relative status we behave assertively or submissively. We establish social and political alliances to gain access to mates and money. We cooperate with unrelated individuals when we benefit directly (through reciprocation) or indirectly (through reputation and other effects). When resources are limited and competition is harsh, we try to hurt our competitors without having to pay a price for it. We establish bonds when we need them, we develop strategies for testing their strength, and we are sensitive to changes in the ratios of costs and benefits of our partnerships. We choose mates, political allies, and business partners through complex negotiations in a biological market in which the value of individuals and their resources fluctuates according to the laws of supply and demand. Throughout many of these transactions and relationships, we behave according to models from game theory and other branches of economics and evolutionary biology. Again, there is a great deal of individual and cultural variation in human behavior. But the average behavior of human beings in many social situations is adaptive and highly predictable. And in many cases, behavioral variation among individuals is adaptive and highly predictable too.

Some of the social problems confronted by modern humans are also present in other organisms, and their adaptive solutions are similar to ours. Nepotistic behavior is widespread among honeybees and ants, fish play tit-for-tat strategies of cooperation, birds form pair-bonds to jointly rear their offspring, dominance hierarchies and ranks are widespread in birds and mammals, and complex strategies of alliance formation are seen in primates, hyenas, and dolphins. In many cases, different groups of animals confronting similar environmental problems come up with similar adaptive solutions independently of one another, a phenomenon that evolutionary biologists call *convergent evolution*. When solutions to a particular problem are limited, natural selection sometimes comes up with the same solution over and over again, even in organisms that are distantly related, such as fish and people. In some cases, the solutions are only superficially similar: both fish and people play tit-for-tat strategies of cooperation, but the cognitive mechanisms that these species use to implement the strategy are likely to be very different. People can think about the future and the consequences of their actions, and they can predict others' responses to their own behavior. Fish probably just have hardwired brain mechanisms that make them do the right thing. In some situations, however, the hardwired mechanisms work so well that there is no need to replace them with more sophisticated cognitive ones. Evolution has to overcome considerable friction to transform a "hardwired" behavioral strategy into a "cognitive" one.[12] Even in animals with big brains, including humans, there will be selection against the use of complicated strategies that are costly in time and cognitive processing power if a simple rule of thumb can work just as well.

The Phylogenetic History of Human Social Behavior

All organisms on the planet, including humans, represent a combination of traits that are new and unique to the species because they evolved recently and traits that are ancient because they were inherited from ancestors. When particular behavioral solutions to environmental problems work well for certain organisms, these solutions can last for a long evolutionary time. When species evolve and give rise to new species, the descendants inherit not only anatomical and physiological adaptations from their ancestors but some of their behavioral adaptations as well. Some behavioral programs that humans use to solve particular social problems are similar to the programs used by other primates, not because we independently came up with the same solution to the problem, but because we and the other primates directly inherited these programs from our common ancestors. Some human emotions, such as fear, have a long evolutionary history. We humans didn't invent anything new in the fear business: we inherited the whole package—the emotion, its physiological mechanisms, and its effects on behavior—from our ancestors. That some emotional programs have a long evolutionary history is not a controversial issue. For other programs, however, the issue of history is controversial. Before I get into the controversy, let me backtrack for a second and go over some very basic information about evolution.

Macroevolution is the process through which species change over time and give rise to new species or go extinct. Evolution can be visualized as a branching process in which some branches of a tree grow and produce new branches while others reach a dead end. All organisms on earth that descended from the same microorganism ancestor are thus evolutionarily interrelated. This can be visualized with a phylogenetic tree, such as the tree of life shown in Figure 9.1. A phylogenetic tree is a branching diagram showing the evolutionary re-

Figure 9.1. Tree of life, depicting the evolutionary relationships among all living organisms. Adapted from "A Simplified Family Tree of Life" in Hotton (1968).

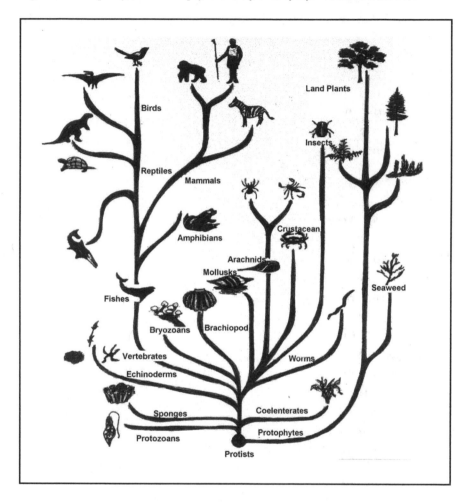

lationships among species or taxonomic groups. These relationships are inferred from similarities and differences in their physical or genetic characteristics. The taxonomic groups joined together in the tree are thought to have descended from a common ancestor. Each node with descendants represents the most recent common ancestor of the descendants, and the edge lengths may be interpreted as estimated time spans between species. Descendant species inherit much of their DNA and the traits coded by the genes from their ancestors. As a result, we can reconstruct the phylogenetic history of traits, such as the presence of mammary glands, by mapping onto a tree and identifying the ancestor in which the trait first appeared. In this case, it is the first mammal that branched off from the vertebrate branch. Some traits can appear for the first time in a species and therefore have no traceable phylogenetic history, or they may look as if they appeared for the first time in a species because information about the ancestors in which the trait evolved has been lost. For instance, language appears to be unique to modern humans. However, it is possible that rudiments of language first evolved in some of our recent ancestors, such as the Australopithecines or other species of the genus *Homo*, which are now extinct. It is possible that language has its own phylogenetic history but that it is difficult to know what it is because our recent ancestors are no longer around.

Many evolutionary psychologists believe that the modern human mind evolved during the Pleistocene, a period that ran from about two and a half million years ago until twelve thousand years before the present time. At this time, our hominid ancestors had already split from the ancestors of chimpanzees, and many of them had already gone extinct. The genus *Homo* made its first appearance during the Pleistocene, and all the fossils that come from this era belong to *Homo erectus* and *Homo sapiens*, which acquired the anatomy of modern humans approximately fifty thousand years ago. Thus, many evolutionary psychologists believe that the modern human mind evolved during the period in which humans acquired their current anatomy, in what they call our "environment of evolutionary adaptedness." Because of this belief, many evolutionary psychologists do not find it useful to try to reconstruct the phylogenetic history of the human mind and behavior by studying other primates, or other animals in general. When *Homo sapiens* developed its big new brain, their argument goes, all bets were off and humans started playing new games with each other, with entirely new behavioral rules. For instance, Tooby and Cosmides, considered by many the founders of modern evolutionary psychology, have stated many times that studying the phylogenetic history of human behavior by comparing it to the behavior of other animals is not useful, or even possible.[13] But others disagree.

I will use an analogy between body anatomy and behavior to explain why I think this position is wrong. The body of *Homo sapiens* might have modified its shape and acquired its current anatomical features during the Pleistocene, but we can nevertheless clearly see two things. First, there are still very strong similarities between the anatomy of modern *Homo sapiens* and that of great apes and other primates; second, from modern human bodies we can still discern elements of a very ancient evolutionary history of human anatomy, which—as nicely illustrated by paleontologist Neil Shubin in his *Your Inner Fish: A Journey into the 3.5-Billion-Year History of the Human Body*—goes all the way back to primitive fish.[14] The same applies to brains and behavior. The mind of *Homo sapiens* might have been modified by natural selection in some important ways during the Pleistocene and acquired all the characteristics of the modern human mind. As we'll see later, it is likely that new cognitive abilities emerged during this period, such as speaking languages, thinking about other people's minds, and making moral judgments. The social problems faced by *Homo sapiens* during the Pleistocene, however, were probably the same problems that its ancestors and other primates had already been dealing with for millions of years. Sure, at some point early humans started talking about these problems over dinner—an evolutionarily novel twist—but I doubt that these dinner conversations led to completely new solutions. We deal with the same problems today, and we still use the same behavioral solutions we inherited from our ancestors.

Behavior has a phylogenetic history, which means that simple behavior patterns as well as complex social strategies are inherited from ancestor species and maintained in new species, sometimes with only minor modifications. Evolutionary biologists call the tendency for descendant species to retain the characteristics of their immediate ancestor *phylogenetic inertia*. When morphological, physiological, behavioral, or psychological traits are similar in different species owing to inheritance from a common ancestor, these traits are called *homologous*. As it turns out, when it comes to social behavior, humans still resemble not only the great apes and other primates but other animals as well, so that in our behavior we can still see phylogenetic traces of our "inner fish" or even our "inner insect." For both anatomy and behavior, the traits that are most likely to be preserved across many branches of the phylogenetic tree are traits that evolved by natural selection and that play an important role in organisms' survival and reproduction; therefore, many adaptations (for example, the emotion of fear) are likely to be homologous across species. Phylogenetic inertia doesn't necessarily hold back the action of natural selection; some of the traits with high inertia are likely to be adaptive.

Some of the reasons why evolutionary psychologists are skeptical about studying the phylogenetic history of human behavior are shared by evolutionary biologists, and this skepticism applies not only to human behavior but to behavior in general. Although the skeptics recognize that behavior evolves by natural selection just as the body does, they maintain that we can't study the phylogenetic history of behavior because behavior is "special" and qualitatively different from the body. One way in which behavior is special is that it is too labile, too variable, too susceptible to influences of the environment to be the subject of phylogenetic analyses. Our body too is affected by the environment in which we live, but not as much as our behavior is. Another difference between bones (or any other part of the body) and behavior is that bones are considered a "structure" while behavior is considered the "functional" product of a structure, the brain. In other words, the brain is a structure equivalent to a bone, and behavior is what the brain does, which would be equivalent to the way a bone moves.

This distinction between structure and function matters because, according to some evolutionary biologists, we can study homologies for structural traits but not for functional ones; for example, we can look at homologies between the legs and brains of animals of different species, but not homologies between the ways these animals walk or behave. A related, but less radical, objection is that two functional traits can be considered homologous only if they are produced by the same structure. For example, rhesus monkey sleep and human sleep could be considered homologous only if sleep is produced and controlled by the same regions of the brain in the two species.

In a rebuttal to these objections, two American primatologists, Drew Rendall and Anthony Di Fiore, have convincingly argued that behavior is not too labile or variable to preclude phylogenetic analyses; that homologies can be established for both functional and structural traits; and that behavioral traits can be homologous even if they originate from different areas of the brain.[15] They thus conclude that there is nothing "special" about behavior from an evolutionary standpoint. And there are data to back up this conclusion: research has shown that similar adaptive behavioral traits shown by different species are as likely to be homologous as morphological traits are. For example, in the early 1990s biologists Alan de Queiroz and Peter Wimberger conducted phylogenetic analyses of morphological and behavioral characteristics in a wide range of animals, including insects, fish, frogs, reptiles, and birds.[16] The morphological characteristics included body size, the size and shape of bones, and other physical characteristics of the animals. The behavioral characteristics ranged from simple stereotyped movements to complex social behaviors (for example, courtship behavior, territoriality, or parental care). They concluded that behavioral traits are as likely to be homologous across different organisms as morphological characters are—as some biologists have known, or suspected, for a long time.

The study of the phylogeny of behavior started in the first half of the twentieth century, gained momentum in the middle of the century, then was virtually abandoned. Now, in the twenty-first century, it is coming back with a vengeance: phylogenetic studies are one of the fastest-growing areas of animal behavioral research. European biologists Konrad Lorenz, Niko Tinbergen, and Karl von Fritsch were awarded the Nobel Prize in Medicine in 1973 for their contribution to the birth of a new scientific discipline, *ethology*, often described as the study of the biological bases of behavior. A major focus of early ethological research was the phylogenetic history of behavior.

Trained in comparative anatomy and systematics, Lorenz and other European ethologists believed that certain sequences of animal movements are as reliable in identifying species as closely or distantly related as any of the morphological characteristics used in comparative anatomy. Two of their predecessors, biologists Charles Whitman (1842–1910) and Oscar Heinroth (1871–1945), had suggested half a century before Lorenz and Tinbergen that the concept of homology was equally applicable to morphological characteristics and behavior. Whitman and Heinroth focused on one particular aspect of behavior: motor patterns. For example, Heinroth argued that the movements involved in yawning and self-scratching are probably homologous among many vertebrate animals. Lorenz and Tinbergen later expanded this type of research by showing that movements in the courtship displays with which male ducks and gulls attract females are very similar in closely related species as a result of inheritance of behavior from common ancestors.[17]

The approach of comparing behavior in closely and distantly related species to find out whether behavior shows evidence of genetic transmission from ancestor to descendant species can be applied to the study of not only movement patterns but also the complex social behaviors involved in mating, parenting, attachment, cooperation, aggression, and submission and defense. These complex behaviors are unlikely to originate from scratch in any single species. In fact, more and more studies encompassing insects, lizards, frogs, birds, and mammals are discovering phylogenetic continuities across different species in a wide range of complex social behaviors. An early example comes from a comparative study of rodents conducted by American biologist John Eisenberg in the 1960s.[18] Eisenberg studied the social systems of kangaroo rats, pocket mice, and related species of rodents and found that each species' social system was better explained by the phylogenetic history of the species—that is, the type of social system possessed by its ancestors and its closest relatives—than by the characteristics of the environment in which the species lived. Studies of social organization in different species of iguana lizards have found that the phylogenetic tree of these lizards explains social traits, such as the presence or absence of male territoriality and male dominance hierarchies, better than it explains morphological traits (for example, differences in body size between males and females).[19]

Similar discoveries have been made in nonhuman primates. In the mid-1970s, two biologists, John Spuhler and Lynn Jorde, classified twenty-one different primate species on the basis of nineteen behavioral characteristics and found that the occurrence of particular behaviors in the various species was explained equally by the characteristics of the environment in which they lived and by the species' position on the primate phylogenetic tree: in other words, a species had inherited particular behaviors from its ancestor.[20] Twenty years later, Di Fiore and Rendall conducted a similar study: they classified sixty-five primate species on the basis of thirty-four aspects of their social organization, such as migration tendencies, grouping, community structure, mating patterns, social relations within and between the sexes, and reproductive investment.[21] They discovered that many aspects of female social behavior, such as the tendency to form dominance hierarchies and aggressive coalitions and the tendency to groom their female relatives, were extremely uniform among Old World monkeys. Essentially, most species of macaques, baboons, langurs, and other Old World monkeys have similar social systems and general patterns of social behavior because these patterns were inherited as a "package" from their ancestors and maintained relatively unmodified during the subsequent evolution and diversification of these monkey species. This finding demonstrates that even very general patterns of behavior associated with social organization can be conserved over large evolutionary time scales. Since many species of Old World monkeys are now quite different from each other in size, physical appearance, and other aspects of their biology, these studies also suggest that behavioral traits may be even more resistant than morphological traits to the evolutionary changes that occur along with species diversification and ecological adaptation to new environments. Another implication is that lacking a knowledge of a species' phylogenetic history may make it difficult for researchers to understand why the species exhibits a particular social system or a particular pattern of complex social behavior. By the same token, it is difficult to understand the evolutionary basis of human social behavior without some knowledge of the social behavior of other primates, and particularly the social behavior of the species closely related to us.

Many aspects of human behavior and cognition were probably inherited from our mammalian ancestors. The probability that two species share similar behaviors owing to common descent is higher the closer the phylogenetic relationship between the species is. The great and lesser apes are, along with the Old World monkeys, the animals that are phylogenetically closest to humans. Therefore, human behavior is more likely to be homologous to the behavior of these primates than to the behavior of other animals. Let's look at the example of the smile, a universal human facial expression reported in all cultures around the globe. In his 1872 book *The Expression of the Emotions in Man and Animals*, Darwin suggested that many human facial expressions evolved from those of our primate ancestors, and primatologists have long believed that the primate "bared-teeth display" and the human smile are homologous, which means that rhesus macaques, chimpanzees, and human beings all inherited this behavior pattern from the ancestor these three species had in common about 25 million years ago.[22]

As mentioned in Chapters 1 and 2, in rhesus macaques and chimpanzees the bared-teeth display is mainly used as a submissive signal. Also known as a "fear grin" or "fear grimace," it is often displayed by low-ranking individuals who have been attacked, threatened, or sometimes simply approached by a high-ranking individual. The signal reflects a combination of fear and a plea for mercy ("don't attack me!"). Rhesus macaques and other species of Old World monkeys and apes also use this signal during friendly interactions, such as when an adult male approaches a female and invites her to mate, or when a mother encourages her baby to walk and follow her. Human beings use the smile mainly for friendly purposes, but its submissive component still exists, as when we nervously smile at our boss. The primate bared-teeth display and the human smile are produced by contraction

Figure 9.2. Left: *The bared-teeth display shown by a rhesus macaque.* Center: *The bared-teeth display shown by a chimpanzee. Left and center photos courtesy of Dr. Lisa Parr.* Right: *The human smile. Photo by Dario Maestripieri.*

of some of the same facial muscles and have similar submissive and friendly functions. Although it would be difficult to provide conclusive and incontrovertible evidence that they are homologous, this explanation seems more reasonable and more likely to be correct than the alternative one, according to which the human smile is an evolutionary novelty that does not have a long phylogenetic history.

A smile is a relatively simple behavior pattern, and an easy case can be made that other simple behavior patterns, such as yawning, have an even longer evolutionary history, given that they are ubiquitous not just in primates but in all mammals. A good example of a more complex program—one with emotional, cognitive, and behavioral components—that humans probably inherited as a package from their primate ancestors is the infant attachment system I discussed in Chapter 6. Many primate infants are dependent on their mother for nutrition, thermoregulation, and protection; as a consequence, they need to be carried all the time or kept in close proximity to their mother. The problem is that if infants become separated and lost from their caregivers, they starve or freeze to death or are eaten by a predator. To solve this problem, natural selection came up with the infant attachment system. As mentioned in Chapter 6, the infant attachment system has a set goal—the maintenance of contact with or proximity to the mother—and specific activating and terminating conditions. The attachment system is activated when the infant is separated from the mother, and it's terminated when contact or proximity is achieved. The infant attachment system has three defining features, or what evolutionary psychologists call *design features*: separation anxiety and fear of strangers; use of the mother as a "safe haven" for protection; and use of the mother as a "secure base" for exploration.

A humanlike infant attachment system with the same set goal, the same activating and terminating conditions, the same design features, and many of the same behaviors, such as infant crying and following, is almost ubiquitous among the Old World monkeys and the apes and absent or rare among the prosimians and the New World monkeys. This suggests that the infant attachment system is an adaptation whose history can be tracked in the evolution of the Primate order. Although there may be slight differences among species in some of the physiological mechanisms or cognitive processes underlying the regulation of attachment processes, the attachment system as a whole shows evidence of homology across primate species.[23]

The Recent Past

The amazing growth in the size and complexity of the human brain that occurred after our hominid ancestors split from those of the other apes was accompanied by the acquisition of many new "mental powers." We developed the ability to think about ourselves, about events happening in the future, and about concepts such as life, death, and the immensity of the universe. The evolution of language gave a big boost to both our ability to think and our ability to communicate with others. Many new mental abilities evolved to solve the problems of a complex social life, and these new abilities, in turn, made our social life even more complex. If it is true, as primatologist Marc Hauser has argued in his book *Moral Minds*, that our ability to think about the moral value of our actions and to make moral judgments about our own actions and those of others evolved by natural selection, this process must have occurred relatively recently.[24] Although the social contract view of morality has validity, it is likely that complex new emotions evolved to support morality, such as shame (subjective penalty for norm violation), guilt (subjective penalty for violation of expectations), pride (subjective payoff for norm adherence), moral outrage (anger that occurs when norm violations by others are experienced as if they were transgressions against the self), and contempt (long-lasting condemnation of others who have transgressed norms or violated expectations). These emotions appear to be evolutionarily recent and unique to humans. Consistent with Noë's suggestion (mentioned in Chapter 8) that team-playing may have played an important role in human social evolution,[25] evolutionary psychologists Fessley and Haley also suggest that "corporate" emotions such as admiration and elevation—which reward individuals who are team players, who behave altruistically, and whose actions benefit others or society as a whole—may also have recently evolved in the human lineage.[26] Finally, pressure to form and maintain heterosexual pair-bonds for joint child-rearing and to invest in our children for the rest of our lives probably favored the evolution of romantic love and parental love to a degree that is not observed in other animals.

We perceive the activity of our own mind as a fully subjective experience. Although we have no direct access to other people's minds, we imagine that these minds exist, that they work like ours, and that they guide other people's behavior through beliefs, desires, knowledge, or ignorance. The ability to think about other people's minds—what psychologists call "a theory of mind"—was accompanied by the ability to engage in complex forms of imitation, teaching, and deception. We also became capable of making complex predictions about the future behavior of other individuals, making complex mental calculations of the costs and benefits of our actions and those of other individuals, and anticipating the behavior of others the way chess players anticipate the moves of their adversaries and their own moves in response to them.

Although our new language abilities, our new ability to think and act morally, our new emotions and feelings, and our new cognitive abilities could have revolutionized the way we negotiate our personal and business relationships as human beings, in reality this hasn't happened—perhaps in much the same way that new technologies such as TV, radios, and computers have not replaced books. Our new mental powers have not replaced the psychological and behavioral predispositions we have inherited from our primate ancestors. Rather, they stand side by side in our brains the way books and the iPad lie side by side on our desk. We still use adaptive behavioral solutions to social problems that are similar to those evolved by distantly related animals such as fish or birds. Since many of our social problems are ancient, we routinely use ancient solutions to solve them. Our intellectual education gives us the opportunity to engage in amazing artistic, scientific, and scholarly achievements. With our moral or religious education and our love for our family, our friends, and the members of our country or any other group to which we belong—which may include the whole human race or all living organisms—we are motivated to engage in virtuous behavior that is valued, praised, and admired by our fellow human beings. And yet, despite how intellectually accomplished or morally virtuous we are, we continue to solve everyday social problems by resorting to the ancient emotional, cognitive, and behavioral algorithms that crowd our minds, often letting this automatic pilot help us navigate through the difficult and dangerous, but always fascinating, waters of human social affairs.

EPILOGUE

At 9:00 A.M. on September 18, 2010, a Saturday, I received an email from someone named Mitchell Heisman. Sent from a Hotmail account, the message was addressed to my University of Chicago email account as well as to a few hundred other email addresses, many of which had the "-edu" suffix of American universities, and to other email addresses that appeared to belong to newspaper reporters and government officials. The subject line read "suicide note"; attached to the email was a large PDF. About one hour after I received the email, a thirty-five-year-old man named Mitchell Heisman shot himself to death on the steps of Harvard University's Memorial Church—while a Yom Kippur service was going on inside.

I had never heard of Mitchell Heisman and suspect that, with a few exceptions, none of the other people who received the same email knew him either. Although I never open attachments from strangers, that day I did. The file contained a book manuscript, 1,905 pages long, including a 20-page bibliography. I couldn't help but notice that my book *Macachiavellian Intelligence: How Rhesus Macaques and Humans Have Conquered the World* was in the list of references. I then discovered that Heisman discussed the book in great detail and in fact had pulled many quotes from it. I was shaken by the possibility that something I wrote could have somehow contributed to someone's decision to end his own life. I started reading the manuscript from the beginning.

I quickly gathered that the book was an intellectual inquiry into human nature, the history of philosophical and political thought in Western societies, the role of science and objectivity in understanding reality, and, ultimately, the meaning of life. The conclusions of the inquiry were not too uplifting.

A couple of hours later, in the midst of my reading, a person named Jared Nathanson replied-all to Mitchell Heisman's note. His email read as follows:

> *Mitchell,*
>
> *You can't argue the meaning or meaningless qualities of life or Democracy in absolutes. The human mind goes crazy within such attempts of discursive explanation. The idea that anything is meaningless because you cannot fathom meaning or because you can construct an elaborate thesis only exists within your own context.*
>
> *Life is. We are. Whatever the meaning, the brutality, or beauty, we exist. To say that there should be a meaning or that meaning isn't there is in itself, a meaningless task. We exist. The alternative is not to exist. For whatever finite moments of joy and sadness and perhaps intellectual curiosity we experience, we will never solve it all and we (meaning you in this instance) will never come up with a cohesive answer to why one philosophical journey is inherently flawed or not. Like most scholars, you are caught up in the rhetoric of your self-made castle of words.*
>
> *What I do know is that once life is over, it is very possible that whatever argument you are making ends with the static confines of your thesis and that your drama in presentation will not cause it to go on forever. The vanity of the suicidal person, the self-centered nature causes the need to exit with a bang, to be noticed by the very people who you probably pity and judge as the unwashed masses.*
>
> *The only way that you can prove your thesis or allow it to live past these few moments and perhaps a tabloid moment on television is to live. Live to argue, live to discuss, live to see if you are right or not. Without you standing by your work, you are abandoning the baby, leaving something you obviously put great time into, simply so you can inaugurate its unique publishing. But your work will die, and not because of a great conspiracy that you think will stifle it, but because the drama of your actions will steal the attention from your work. You will*

take a work of intellectual labor and reduce it to an episode of the Jersey
Shore. That is a sad thing for an intelligent man to do.

If you decide to stay on the planet, I'd be happy to read your work
thoroughly and discuss.

Thanks,

—Jared

I immediately Googled Jared Nathanson and discovered that he was the lead singer of a Boston music band called the Heartsleeves. (On their website, www.heartsleeves.com, they describe their music as "Neo Eclectic, Soul Reflected, Sounds of Real Life!") I decided to email him and ask him what was going on. Jared replied and, among other things, explained to me that

> I knew Mitch, though it must be said I didn't recall his name when I received the email or when I responded. I didn't mean to send my response to everyone, that was unfortunate. As a matter of fact, I wasn't sure at that moment that it wasn't some kind of internet scam. I responded because I had the feeling that I might know this person or at the least, I understood his plight, trying to find meaning in a world without simple moral narratives to guide us.

A few days later, Jared sent me another email with a link to an article about Mitchell Heisman that had just appeared on the online edition of the *Boston Globe*.[1] In the article, reporter David Abel provided some biographical information about Mitchell, his family and upbringing, and his living conditions prior to his death. Family members and friends remembered him as a gregarious child who had grown introverted after his father died of a heart attack when Mitchell was only twelve years old. He had studied psychology at the University at Albany in New York, where he generally avoided people and spent much of his time reading. After college, Mitchell worked at bookstores and accumulated a library of thousands of books. He then started working on his book full-time. He lived alone and survived on microwave meals, chicken wings, and energy bars. To better concentrate for his writing, Mitchell often listened to a constant loop of Bach's "Well-Tempered Clavier" and took Ritalin. On the morning of Yom Kippur, Mitchell showered, shaved, and ate a breakfast of chicken fingers and lentils. He put on a trench coat over a white tuxedo, with white socks and shoes. Then he went to Harvard and shot himself.

Mitchell was on a quest to understand himself and the world around him. As a scholar, he used scientific and logical reasoning to examine and evaluate theories and discoveries produced by biologists, psychologists, historians, philosophers, and other researchers. His inquiry led him to conclude that evolutionary biology provides the most direct answers to questions about the self and human nature. In his book, he argued that many of our emotions, feelings, and thoughts reflect biological predispositions that help us survive and reproduce. He also wrote that many patterns in human history can be understood as the outcome of nepotistic cooperation among members of one's own family or group and competition against members of other groups, and that similar social dynamics occur in other primates as well. When discussing the controversies that followed the publication in 1975 of Edward O. Wilson's book *Sociobiology: The New Synthesis*, Mitchell bluntly stated: "The problem is not that sociobiology does not make sense. The problem is that sociobiology makes too much sense."

Satisfied with the way his scientific reasoning produced the knowledge and explanations he was looking for, Mitchell tried to use the same approach to search for a justification of knowledge-seeking in itself and, ultimately, of his own existence. He wanted to be objective at all costs and eliminate all sources of bias that might cloud his analysis, especially the psychological predispositions toward self-interest, survival and reproduction, and life in general. But after searching for and eliminating all of these subjective biases, he couldn't find any rational justification for knowledge or life. So Mitchell concluded that, taken to its extreme, striving for objectivity ultimately leads to nihilism and rational self-destruction. In his own words: "Life is a prejudice that happens to be talented at perpetuating or replicating itself. To attempt to eliminate this source of bias is to open your mind to death. I cannot fully reconcile my understanding of the world with my existence in it. There is a conflict between the value of objectivity and the facts of my life." He committed suicide as an experiment to demonstrate the incompatibility between "truth" and "life."

When I was still a high school—and later college—student, I was an introvert who spent a lot of time reading books and thinking. Like Mitchell, I developed an interest in scientific and philosophical thinking as a way to try to understand myself and others around me and to answer the eternal questions of who we are, where we come from, and where we are going. I, too, arrived at the belief that evolutionary biology has the most direct answers to these questions, and eventually I became one of those scientists who do "me-search." I was especially fascinated by human behavior and by its parallels with animal behavior. If Mitchell and I had met and discussed our ideas, I would have agreed with many of his analyses of human nature and human history. I would even have accepted his conclusion that "life is a prejudice that happens to be talented at perpetuating or replicating itself." Unlike Mitchell, however, I don't have any problems with that. I find this "prejudice" to be interesting, beautiful, and well worth living.

Mitchell's death is a great loss for his family, friends, and humankind. But ultimately, Mitchell was the only one responsible for his death—not the people he met in his life, not the people whose books he read, and certainly not evolutionary biologists and their explanations of life and human nature. As Jared Nathanson tried to explain in his email to Mitchell that didn't reach him in time, the justification for living one's life doesn't come from the explanation of what life is but from life itself. It is possible that had Mitchell lived a better life, he would have chosen to keep on living.

Evolutionary explanations of the origins and processes of life—as opposed, for example, to the religious explanations advanced by creationists—are viewed by critics as being cynical, pessimistic, and depressing because they essentially maintain that life came out of nowhere. It simply emerged from some lucky mix of rocks, gas, and water. Nor is it going anywhere, since evolution does not have an ultimate goal, such as the reaching of complexity or perfection. The "selfish gene" view of natural selection articulated and publicized by evolutionary biologist Richard Dawkins is also accused of being cynical, because it implies that organisms are merely vehicles for genes and that genes are preprogrammed to advance their own cause in competition with other genes for survival and propagation. Finally, evolutionary explanations of human social behavior are often labeled as cynical, because adaptive behavior is viewed as the product of cost-benefit ratios that are advantageous to individuals and their genes, often at the expense of others.

These critics make the same mistake Mitchell made. They misunderstand what science is and what it does. Science produces knowledge and explanations, not philosophical, moral, or religious justifications. Evolutionary biology is a scientific discipline; its job is to help us understand what life is and how it works. Evolutionary biology has no business telling us whether or not life is worth living and why. In my view, whether or not life is worth living depends on the quality of one's own life, which in turn may depend on one's physical and mental health, one's happiness or unhappiness. I like to think that there is a threshold level for quality of life, below which one's life may not be worth living, especially if the prospects for improvement are nonexistent—a rare situation, but one that may nevertheless occur in some unfortunate circumstances.

Science can improve the quality of our lives in many ways, such as through medicine or useful technologies. Knowledge produced by scientific research can also empower us, increasing our capacity to control our lives and accomplish our goals, and knowledge of the natural world and of all living organisms may lead us to an appreciation of the beauty in nature. However, science does not and cannot provide philosophical reasons that justify why life is worth living or why knowledge is worth pursuing.

Nature is neither good nor bad; therefore, explanations of the natural world cannot be optimistic or pessimistic. Nor is human nature simply good or bad, and "rational" explanations of human behavior such as those provided by evolutionary biologists or economists cannot be optimistic or pessimistic, uplifting or depressing, hopeful or cynical. People search for happiness in many different ways throughout their lives, and being happy may include feeling good about oneself and having a positive outlook on life and the world. Although knowledge of oneself and the world may be instrumental in the pursuit of happiness, the truth is that there is no correlation between knowledge and happiness. People who are clueless about themselves and the world can nevertheless be very happy. Understanding ourselves, life, and the world we live in can be useful, and indeed a lot of fun, but I don't think that there is an overall justification for the process of seeking knowledge of the kind that Mitchell Heisman was looking for, just as there is no overall justification or "meaning" of life.

Instead of realizing that the only way to become happier is to improve the quality of one's life, many people need reassurance that all stories have a happy ending, that humans are fundamentally good, that bad things don't happen to good people, or that there is a supernatural being who takes care of us and makes sure everything is okay. If a scientist attempts to explain nature or human nature but doesn't provide a positive message that makes people feel better about themselves, some of them will simply shoot the messenger. Scientists who communicate often with the general public have learned that their audiences expect a positive message from their science, the way many moviegoers expect that a film will have a happy ending. Movies with a happy ending are probably more successful than movies with an unhappy ending or no clear ending at all, and scientific books with a "positive message" probably sell many more copies than books that don't have one.

The notion that human beings have evolved from primate ancestors and that their behavior shares many similarities with the behavior of extant monkeys and apes is itself neither good nor bad. The scientists who attempt to educate the general public about our evolutionary ancestry and close genetic relatedness to other primates can convey a positive message if they maintain that these primates are fundamentally good-natured. Conversely, people may perceive that a message is bad if our close primate relatives are described as selfish, Machiavellian, or murderous creatures. Obviously, primates are neither good nor bad, neither good-natured nor evil. Therefore, recognizing and understanding our evolutionary kinship to other primates carries no implications whatsoever in terms of establishing the inherent goodness or badness of human nature. That would be beside the point. What our knowledge of the "games primates play" does do is help us understand that human nature exists, explain what it is, and help us learn how it works. And that's all we can really ask for.

ACKNOWLEDGMENTS

I would like to thank my agent, Esmond Harmsworth, who encouraged me to write this book, helped me with the proposal, and read and commented on the entire manuscript. I also would like to thank TJ Kelleher and Tisse Takagi at Basic Books for doing so much to improve my manuscript. Thanks also to all the friends and colleagues who read and gave comments on some of the chapters, and in particular to Jennifer Beshel, who read and edited all of them. Finally, special thanks to Sian for her support and encouragement during the writing of this book, and for reading everything I wrote and making it better.

NOTES

INTRODUCTION

1. Cacioppo and Patrick (2009).

2. Berne (1964).

3. One of the first systematic studies of social relationships in nonhuman primates is Robert Hinde's *Primate Social Relationships* (1983). Hinde also wrote a book about human relationships from a comparative perspective (Hinde 1997).

4. Gould (1990).

5. For one of the first systematic discussions of this trend, see Jerison (1973).

6. The competitive chimpanzee society is described in detail in Jane Goodall's book *The Chimpanzees of Gombe* (1986). The rhesus macaque is another primate species that lives in a highly competitive society (see Maestripieri 2007).

7. For a comprehensive discussion of sex differences in mate-seeking and mate attractiveness, see the book by evolutionary psychologist David Buss, *The Evolution of Desire* (1994).

8. Levitt and Dubner (2005).

9. Wrangham and Peterson (1996).

10. An excerpt from the next chapter was published as "Op-Ed: Why the Elevator Floor Is So Interesting," in *Wired*, May 27, 2009, available at: http://www.wired.com/wiredscience/2009/05/ftf-mastripieri.

CHAPTER 1: DILEMMAS IN THE ELEVATOR

1. Hall (1966).

2. For a discussion of the causes of aggression, including spatial proximity, in primates, see Higley (2003).

3. For more information about rhesus macaques, see Maestripieri (2007).

4. For more information about the bared-teeth display and the relaxing properties of grooming, see Maestripieri (1996) and Schino et al. (1988).

5. Mike Tomasello and his colleagues have written extensively about the importance of understanding and sharing goals and intentions and the dif-

ferences between humans and other primates in their ability to engage in these processes; see, for example, Tomasello et al. (2005).

6. Taylor (2002).

7. Part of the experiment and its results are described in Schino et al. (1990).

8. The Prisoner's Dilemma and the dynamics of cooperation are discussed extensively in Robert Axelrod's book *The Evolution of Cooperation* (1984), as well as in Dugatkin (1997).

Chapter 2: The Obsession with Dominance

1. For a description of the dynamics of grooming between male chimpanzees, see Simpson (1973).

2. Trivers (1985) discusses the psychological and behavioral tactics used by parents and children when parent-offspring conflict arises.

3. Chikazawa et al. (1979).

4. Mock (2004).

5. For an example of research on dominance in children, see Pettit et al. (1990).

6. Packer (1977); Sapolsky (1992).

7. For a reissued edition of the English translation of Canetti's 1935 novel, see Canetti (1984).

8. I extensively discuss primate dominance in my book *Macachiavellian Intelligence* (Maestripieri 2007).

9. Conniff (2003, 2005).

10. Wilson (1975).

11. On the debate about primate dominance, see Hinde (1972), Rowell (1974), Wilson (1975), Altmann (1981), Bernstein (1981), Hinde and Datta (1981), de Waal (1986), and Lewis (2002). Figure 2.2 was modified after Aureli and Whiten (2003).

12. John Maynard-Smith's pioneering applications of game theory to animal conflicts and dominance are presented in Maynard-Smith and Price (1973) and Maynard-Smith and Parker (1976); for a more general discussion of the use of game theory in evolutionary biology and animal behavior, see Maynard-Smith (1982). For a discussion of behavioral displays and other signals, see Maynard-Smith and Harper (2003).

13. The relationship between dominance and neuroendocrine variables is discussed in Cummins (2005).

Chapter 3: We Are All Mafiosi

1. D. Carlucci, G. Di Feo, and G. Foschini, "La Mafia dei baroni," *L'Espresso*, January 27, 2007, available at: http://espresso.repubblica.it /dettaglio/la-mafia-dei-baroni/1481927.

2. Bellow (2003).

3. The biological basis of nepotism, kin selection, was first formally discovered in the 1960s by British evolutionary biologist William Hamilton (1964a, 1964b).

4. I extensively discuss the nepotism of rhesus macaques in my book *Macachiavellian Intelligence* (Maestripieri 2007).

5. Bellow (2003) illustrates with many examples the ways in which people have extended the status of kin to strangers, and even to the gods, throughout human history.

6. The story involving Prof. Ezio Capizzano made the headlines of several Italian newspapers and online news sites in December 2001. Professor Capizzano was later acquitted of all charges and wrote a memoir detailing his sexual encounters with students.

7. The empirical test of the "infinite monkey theorem" at the Paignton Zoo in England was conducted in 2003 and reported by various online news sites, including the BBC News ("No Words to Describe Monkeys' Play," May 9, 2003, available at: http://news.bbc.co.uk/2/hi/3013959.stm).

8. Zingales (2012).

Chapter 4: Climbing the Ladder

1. For a detailed discussion of the rhesus macaque population on Cayo Santiago, see my book *Macachiavellian Intelligence* (Maestripieri 2007). For studies of male migration in macaques, see Wheatley (1982), van Noordwijk and van Schaik (1985, 1988), and Dubuc et al. (2011).

2. For these and other examples of winner-take-all markets, see Frank (1996).

3. Van Schaik, Pandit, and Vogel (2006).

4. Manson (1998).

5. Higham's observations are reported in Higham and Maestripieri (2010).

6. Van Schaik et al. (2006).

7. Betzig (1986).

Chapter 5: Cooperate in the Spotlight, Compete in the Dark

1. Bateson, Nettle, and Roberts (2006).

2. Haley and Fessler (2005).

3. The biological predispositions to respond to eyes and eye gaze direction in animals and humans are discussed by Burnham and Hare (2007); see also Emery (2000).

4. For a study of reputation effects on generosity in a Dictator Game, see Servátka (2010).

5. Bshary and Grutter (2006).

6. Hardin (1968).

7. See Nowak and Sigmund (1998) for a study of indirect reciprocity through image scoring.

8. Andreoni and Petrie (2004); Rege and Telle (2004).

9. Burnham and Hare (2007).

10. Milinski, Semmann, and Krambeck (2002a); Semman, Krambeck, and Milinski (2004).

11. Andreoni and Petrie (2004); Milinski, Semmann, and Krambeck (2002b).

12. Alexander (1987), p. 100.

13. Piazza and Bering (2008).

14. Dunbar (1998).

15. Trivers (1971).

16. Studies showing punishment of defectors are summarized by Fehr and Fischbacher (2004); see also Egas and Riedl (2008).

17. Hauser (1992).

18. Further information about the 1977 blackout in New York City can be found at "New York City Blackout of 1977," Wikipedia, available at: http://en.wikipedia.org/wiki/New_York_City_blackout_of_1977.

19. "Professor Said to Be Charged After 3 Are Killed in Alabama," *New York Times*, February 10, 2010, available at: http://www.nytimes.com/2010/02/13/us/13alabama.html.

20. Rothwell and Martyn (2000).

21. Many studies have found evidence of nepotistic, gender, and other biases in single-blind anonymous peer reviews, such as Lloyd (1990), Wenneras and Wold (1997), Link (1998), and Budden et al. (2008). For an alternative view of sex discrimination in the peer review system, see Ceci and Williams (2011).

22. The archive is available at: www2.uah.es/jmc; see also Campanario (1998).

CHAPTER 6: THE ECONOMICS AND EVOLUTIONARY BIOLOGY OF LOVE

1. Becker (1981).

2. Frank (1988).

3. Ibid., p. 211.

4. Morris's 1996 book *Partners in Power: The Clintons and Their America* makes the case that Bill and Hillary Clinton have a strong "business" partnership.

5. Rostand (2003).

6. David Buss's 1994 book *The Evolution of Desire* provides a comprehensive discussion of sexual desire and sexual attraction.

7. See Fisher, Aron, and Brown (2005).

8. See Knott and Kahlenberg (2010) for orangutans and Tardif, Carson, and Gangaware (1990) for tamarins.

9. Fraley, Brumbaugh, and Marks (2005).

10. Konner and Worthman (1980).

11. Fisher (1989, 2004).

12. The results of the ISTAT study were reported in an article in the Italian newspaper *Corriere della Sera*, July 22, 2010.

13. Eastwick (2009).

14. Fraley and Shaver (1998).

15. Bowlby (1969); Maestripieri (2003).

16. For a popularized account of research on adult attachment, see *Attached: The New Science of Adult Attachment and How It Can Help You Find— and Keep—Love* (Levine and Heller 2010).

17. For research on testosterone and romantic relationships, see Ellison and Gray (2009) and Maestripieri et al. (2010). For the effects of fatherhood on testosterone, see Gettler et al. (2011).

18. Hazan and Zeifman (1999).

CHAPTER 7: TESTING THE BOND

1. Although there is some controversy over this issue, according to some dictionaries the words *testicle* and *testify* both derive from the Latin word *testis*, which means "witness."

2. Packer (1977).

3. Whitham and Maestripieri (2003).

4. Smuts and Watanabe (1990); Smuts (2002).

5. Zahavi (1977).

6. Zahavi and Zahavi (1997).

7. The use of infants by male baboons during fights is extensively discussed by Stein (1984).

8. Miller (2009).

9. For further discussion of the parallels between the Handicap Principle and some principles of economics, see Bowles and Hammerstein (2003).

10. The story of this controversy is discussed in Zahavi (2003).

11. Skinner (1957); Chomsky (1959).

12. Wilson (1975).

13. Zahavi (2003), p. 862.

14. This work is described in Zahavi and Zahavi (1997).

15. Perry et al. (2003).

16. Michael Balter's article about Susan Perry's conference presentation, "Probing Culture's Secrets: From Capuchin Monkeys to Children," appeared in *Science*, July 16, 2010, pp. 266–267.

17. Manson (1999).
18. Smith et al. (2011).
19. Creel (1997).
20. Smuts (2002).
21. Zahavi and Zahavi (1997).
22. Buss (2000), p. 208.
23. Zahavi and Zahavi (1997).

CHAPTER 8: SHOPPING FOR PARTNERS IN THE BIOLOGICAL MARKET

1. An early synthesis of research on biological markets is provided in *Economics in Nature: Social Dilemmas, Mate Choice, and Biological Markets* (Noë, van Hooff, and Hammerstein 2001). Noë's chapter in this book (Noë 2001), along with earlier papers by Noë and Hammerstein (1994, 1995), provide much of the conceptual background for this research approach as well as for this chapter.
2. Gangestad and Thornhill (1997).
3. Buss (1994), p. 93.
4. Becker (1981).
5. Grossbard-Shechtman (1993).
6. Pawlowski and Dunbar (1999).
7. Kurzban and Weeden (2005, 2007).
8. Gumert (2007).
9. Metz, Klump, and Friedl (2007).
10. Vahed (1998); Fromhage and Schneider (2005); Noë (2001).
11. Seyfarth (1976, 1977).
12. Smith, Memenis, and Holekamp (2007).
13. Fruteau et al. (2009).
14. Studies of the ants-butterfly larvae market are summarized and discussed in Noë (2001).
15. Bshary (2001); Bshary and Noë (2003); Bshary and Grutter (2006); Bshary and Côté (2008).
16. Adam (2010).
17. As an example of these studies, see Hauk (2001).
18. Just My Best Publishing Company, http://www.jmbpub.com/interest.htm.
19. Chiang (2010).
20. Noë (2007).

CHAPTER 9: THE EVOLUTION OF HUMAN SOCIAL BEHAVIOR

1. O'Brian (1994).
2. Burkhardt (2005).
3. Hitchens (1997).

4. Dawkins (2009).

5. Dennis Overbye, "Free Will: Now You Have It, Now You Don't," *New York Times*, January 2, 2007, available at: http://www.nytimes.com/2007/01/02/science/02free.html.

6. Wegner (2002).

7. For an introduction to the discipline of evolutionary psychology, I recommend reading the online document "Evolutionary Psychology: A Primer" by Leda Cosmides and John Tooby, 1997, available at: http://www.psych.ucsb.edu/research/cep/primer.html.

8. Tooby and Cosmides (1990); Cosmides and Tooby (2000).

9. Fessler and Haley (2003).

10. Cosmides and Tooby (2005).

11. Gigerenzer, Todd, and ABC Research Group (1999).

12. Noë (2006).

13. Tooby and Cosmides (1989, 1992).

14. Shubin (2008).

15. Rendall and Di Fiore (2006).

16. De Queiroz and Wimberger (1993).

17. The work of early ethologists is discussed in Burkhardt (2005).

18. Eisenberg (1963).

19. Carothers (1984).

20. Spuhler and Jorde (1975).

21. Di Fiore and Rendall (1994).

22. Van Hooff (1972).

23. Maestripieri (2003).

24. Hauser (2006).

25. Noë (2007).

26. Fessler and Haley (2003).

Epilogue

1. David Abel, "What He Left Behind: A 1,905-Page Suicide Note," *Boston Globe*, September 27, 2010, available at: www.boston.com/news/local/massachusetts/articles/2010/09/27/book_details_motives_for_suicide_at_harvard/.

REFERENCES

Adam, T. C. 2010. "Competition Encourages Cooperation: Client Fish Receive Higher-Quality Service When Cleaner Fish Compete." *Animal Behaviour* 79: 1183–1189.

Alexander, R. D. 1987. *The Biology of Moral Systems*. New York: Aldine de Gruyter.

Altmann, S. A. 1981. "Dominance Relationships: The Cheshire Cat's Grin?" *Behavioral and Brain Sciences* 4: 430–431.

Andreoni, J., and R. Petrie. 2004. "Public Goods Experiments Without Confidentiality: A Glimpse into Fundraising." *Journal of Public Economics* 88: 1605–1623.

Aureli, F., and A. Whiten. 2003. "Emotions and Behavioral Flexibility." In *Primate Psychology*, edited by D. Maestripieri, pp. 289–323. Cambridge, MA: Harvard University Press.

Axelrod, R. 1984. *The Evolution of Cooperation*. New York: Basic Books.

Bateson, M., D. Nettle, and G. Roberts. 2006. "Cues of Being Watched Enhance Cooperation in a Real-World Setting." *Biology Letters* 2: 412–414.

Becker, G. 1981. *A Treatise on the Family*. Cambridge, MA: Harvard University Press.

Bellow, A. 2003. *In Praise of Nepotism: A Natural History*. New York: Anchor Books.

Berne, E. 1964. *Games People Play: The Psychology of Human Relationships*. New York: Ballantine Books.

Bernstein, I. S. 1981. "Dominance: The Baby and the Bathwater." *Behavioral and Brain Sciences* 4: 419–429.

Betzig, L. 1986. *Despotism and Differential Reproduction: A Darwinian View of History*. Hawthorne, NY: Aldine de Gruyter.

Bowlby, J. 1969. *Attachment and Loss*. New York: Basic Books.

Bowles, S., and P. Hammerstein. 2003. "Does Market Theory Apply to Biology?" In *Genetic and Cultural Evolution of Cooperation*, edited by P. Hammerstein, pp. 153–165. Cambridge, MA: MIT Press.

Bshary, R. 2001. "The Cleaner Fish Market." In *Economics in Nature*, edited by R. Noë, J.A.R.A.M. van Hooff, and P. Hammerstein, pp. 146–172. Cambridge: Cambridge University Press.

Bshary, R., and I. M. Côté. 2008. "New Perspectives on Marine Cleaning Mutualism." In *Fish Behaviour*, edited by C. Magnhagen, V. A. Braithwaite, E. Forsgren, and B. G. Kapoor, pp. 563–592. Enfield, NH: Science Publishers.

Bshary, R., and A. S. Grutter. 2006. "Image Scoring and Cooperation in a Cleaner Fish Mutualism." *Nature* 441: 975–978.

Bshary, R., and R. Noë. 2003. "Biological Markets: The Ubiquitous Influence of Partner Choice on the Dynamics of Cleaner Fish–Client Reef Fish Interactions." In *Genetic and Cultural Evolution of Cooperation*, edited by P. Hammerstein, pp. 167–184. Cambridge, MA: MIT Press.

Budden, A. E., T. Tregenza, L. W. Aarssen, J. Koricheva, R. Leimu, and C. J. Lortie. 2008. "Double-Blind Review Favours Increased Representation of Female Authors." *Trends in Ecology and Evolution* 23: 4–6.

Burkhardt, R. W. 2005. *Patterns of Behavior: Konrad Lorenz, Niko Tinbergen, and the Founding of Ethology*. Chicago: University of Chicago Press.

Burnham, T. C., and B. Hare. 2007. "Engineering Human Cooperation: Does Involuntary Neural Activation Increase Public Goods Contributions?" *Human Nature* 18: 88–108.

Buss, D. M. 1994. *The Evolution of Desire: Strategies of Human Mating*. New York: Basic Books.

———. 2000. *The Dangerous Passion: Why Jealousy Is as Necessary as Love and Sex*. New York: Free Press.

Cacioppo, J. T., and W. Patrick. 2009. *Loneliness: Human Nature and the Need for Social Connection*. New York: Norton.

Campanario, J. M. 1998. "Have Referees Rejected Some of the Most Cited Articles of All Times?" *Journal of the American Society for Information Science and Technology* 47: 302–310.

Canetti, E. 1984. *Auto-da-Fé*. New York: Farrar, Straus & Giroux.

Carothers, J. H. 1984. "Sexual Selection and Sexual Dimorphism in Some Herbivorous Lizards." *American Naturalist* 124: 244–254.

Ceci, S. J., and W. M. Williams. 2011. "Understanding Current Causes of Women's Underrepresentation in Science." *Proceedings of the National Academy of Sciences USA* 108: 3157–3162.

Chiang, Y. S. 2010. "Self-interested Partner Selection Can Lead to the Emergence of Fairness." *Evolution and Human Behavior* 31: 265–270.

Chikazawa, D., T. P. Gordon, C. A. Bean, and I. S. Bernstein. 1979. "Mother-Daughter Dominance Reversals in Rhesus Monkeys (*Macaca mulatta*)." *Primates* 20: 301–305.

REFERENCES

Chomsky, N. 1959. Review of B. F. Skinner's *Verbal Behavior*. *Language* 35: 26–58.

Conniff, R. 2003. *The Natural History of the Rich: A Field Guide*. New York: Norton.

———. 2005. *The Ape in the Corner Office: How to Make Friends, Win Fights, and Work Smarter by Understanding Human Nature*. New York: Three Rivers Press.

Cosmides, L., and J. Tooby. 2000. "Evolutionary Psychology and the Emotions." In *Handbook of Emotions*, edited by M. Lewis and J. M. Haviland-Jones, 2nd ed., pp. 91–115. New York: Guilford Press.

———. 2005. "Neurocognitive Adaptations Designed for Social Exchange." In *The Handbook of Evolutionary Psychology*, edited by D. M. Buss, pp. 584–627. Hoboken, NJ: Wiley & Sons.

Creel, S. 1997. "Cooperative Hunting and Group Size: Assumptions and Currencies." *Animal Behaviour* 54: 1319–1324.

Cummins, D. 2005. "Dominance, Status, and Social Hierarchies." In *The Handbook of Evolutionary Psychology*, edited by D. M. Buss, pp. 676–697. Hoboken, NJ: John Wiley & Sons.

Dawkins, R. 2009. *The Greatest Show on Earth: The Evidence for Evolution*. New York: Free Press.

De Waal, F.B.M. 1986. "The Integration of Dominance and Social Bonding in Primates." *Quarterly Review of Biology* 61: 459–479.

De Queiroz, A., and P. H. Wimberger. 1993. "The Usefulness of Behavior for Phylogeny Estimation: Levels of Homoplasy in Behavioral and Morphological Characters." *Evolution* 47: 46–60.

Di Fiore, A., and D. Rendall. 1994. "Evolution of Social Organization: A Reappraisal for Primates by Using Phylogenetic Methods." *Proceedings of the National Academy of Sciences USA* 91: 9941–9945.

Dubuc, C., L. Muniz, M. Heistermann, A. Engelhardt, and A. Widdig. 2011. "Testing the Priority-of-Access Model in a Seasonally Breeding Primate Species." *Behavioral Ecology and Sociobiology* 65: 1615–1627.

Dugatkin, L. A. 1997. *Cooperation Among Animals: An Evolutionary Perspective*. Oxford: Oxford University Press.

Dunbar, R.I.M. 1998. *Grooming, Gossip, and the Evolution of Language*. Cambridge, MA: Harvard University Press.

Eastwick, P. W. 2009. "Beyond the Pleistocene: Using Phylogeny and Constraint to Inform the Evolutionary Psychology of Human Mating." *Psychological Bulletin* 135: 794–821.

Egas, M., and A. Riedl. 2008. "The Economics of Altruistic Punishment and the Maintenance of Cooperation." *Proceedings of the Royal Society of London B* 275: 871–878.

Eisenberg, J. F. 1963. *The Behavior of Heteromyid Rodents.* Berkeley: University of California Press.

Ellison, P. T., and P. B. Gray, eds. 2009. *The Endocrinology of Social Relationships.* Cambridge, MA: Harvard University Press.

Emery, N. J. 2000. "The Eyes Have It: The Neuroethology, Function, and Evolution of Social Gaze." *Neuroscience and Biobehavioral Reviews* 24: 581–604.

Fehr, E., and U. Fischbacher. 2004. "Social Norms and Human Cooperation." *Trends in Cognitive Sciences* 8: 185–190.

Fessler, D.M.T., and K. J. Haley. 2003. "The Strategy of Affect: Emotions in Human Cooperation." In *The Genetic and Cultural Evolution of Cooperation*, edited by P. Hammerstein, pp. 7–36. Cambridge, MA: MIT Press.

Fisher, H. 1989. "Evolution of Serial Pairbonding." *American Journal of Physical Anthropology* 78: 331–354.

———. 2004. *Why We Love: The Nature and Chemistry of Romantic Love.* New York: Holt.

Fisher, H., A. Aron, and L. L. Brown. 2005. "Romantic Love: An fMRI Study of a Neural Mechanism for Mate Choice." *Journal of Comparative Neurology* 493: 58–62.

Fraley, R. C., C. C. Brumbaugh, and M. J. Marks. 2005. "The Evolution and Function of Adult Attachment: A Comparative and Phylogenetic Analysis." *Journal of Personality and Social Psychology* 89: 731–746.

Fraley, R. C., and P. R. Shaver. 1998. "Airport Separations: A Naturalistic Study of Adult Attachment Dynamics in Separating Couples." *Journal of Personality and Social Psychology* 75: 1198–1212.

Frank, R. H. 1988. *Passions Within Reason: The Strategic Role of the Emotions.* New York: Norton.

———. 1996. *The Winner-Take-All Society: Why the Few at the Top Get So Much More Than the Rest of Us.* New York: Penguin.

Fromhage, L., and J. M. Schneider. 2005. "Safer Sex with Feeding Females: Sexual Conflict in a Cannibalistic Spider." *Behavioral Ecology* 16: 377–382.

Fruteau, C., B. Voelkl, E. van Damme, and R. Noë. 2009. "Supply and Demand Determine the Market Value of Food Providers in Wild Vervet Monkeys." *Proceedings of the National Academy of Sciences USA* 106: 12007–12012.

Gangestad, S. W., and R. Thornhill. 1997. "The Evolutionary Psychology of Extra-Pair Sex: The Role of Fluctuating Asymmetry." *Evolution and Human Behavior* 18: 69–88.

Gettler, L. T., T. W. McDade, A. B. Feranil, and C. W. Kuzawa. 2011. "Longitudinal Evidence That Fatherhood Decreases Testosterone in Human

Males." *Proceedings of the National Academy of Sciences USA*, 108: 16194–16199.

Gigerenzer, G., P. M. Todd, and ABC Research Group. 1999. *Simple Heuristics That Make Us Smart*. Oxford: Oxford University Press.

Goodall, J. 1986. *The Chimpanzees of Gombe*. Cambridge, MA: Belknap Press of Harvard University.

Gould, S. J. 1990. *Wonderful Life: The Burgess Shale and the Nature of History*. New York: Norton.

Grossbard-Shechtman, S. 1993. *On the Economics of Marriage: A Theory of Marriage, Labor, and Divorce*. Boulder, CO: Westview Press.

Gumert, M. D. 2007. "Payment for Sex in a Macaque Mating Market." *Animal Behaviour* 74: 1655–1667.

Haley, K. J., and D. Fessler. 2005. "Nobody's Watching? Subtle Cues Affect Generosity in an Anonymous Economic Game." *Evolution and Human Behavior* 26: 245–256.

Hall, E. T. 1966. *The Hidden Dimension*. New York: Anchor Books.

Hamilton, W. D. 1964a. "The Genetical Evolution of Social Behaviour. I." *Journal of Theoretical Biology* 7: 1–16.

———. 1964b. "The Genetical Evolution of Social Behaviour. II." *Journal of Theoretical Biology* 7: 17–52.

Hardin, G. 1968. "The Tragedy of the Commons." *Science* 162: 1243–1248.

Hauk, E. 2001. "Leaving the Prison: Permitting Partner Choice and Refusal in Prisoner's Dilemma Games." *Computational Economics* 18: 65–87.

Hauser, M. D. 1992. "Costs of Deception: Cheaters Are Punished in Rhesus Monkeys." *Proceedings of the National Academy of Sciences USA* 89: 12137–12139.

———. 2006. *Moral Minds: How Nature Designed Our Universal Sense of Right and Wrong*. New York: HarperCollins.

Hazan, C., and D. Zeifman. 1999. "Pair Bonds as Attachments: Evaluating the Evidence." *The Handbook of Attachment: Theory, Research, and Clinical Applications*, edited by J. Cassidy and P. R. Shaver, pp. 336–354. New York: Guilford Press.

Higham, J. P., and D. Maestripieri. 2010. "Revolutionary Coalitions in Male Rhesus Macaques." *Behaviour* 147: 1889–1908.

Higley, J. D. 2003. "Aggression." In *Primate Psychology*, edited by D. Maestripieri, pp. 15–40. Cambridge, MA: Harvard University Press.

Hinde, R. A. 1972. "Concepts of Emotion." In *Physiology, Emotion, and Psychosomatic Illness*, Ciba Foundation Symposium 8: pp. 3–13. Amsterdam: Associated Medical Publishing.

———, ed. 1983. *Primate Social Relationships: An Integrated Approach*. London: Psychology Press.

REFERENCES

———. 1997. *Relationships: A Dialectical Perspective*. Oxford: Blackwell.

Hinde, R. A., and S. Datta. 1981. "Dominance: An Intervening Variable." *Behavioral and Brain Sciences* 4: 442.

Hitchens, C. 1997. *The Missionary Position: Mother Teresa in Theory and Practice*. New York: Verso.

Hotton, N., III. 1968. *The Evidence of Evolution*. Smithsonian Series. Washington, DC: American Heritage Publishing Co.

Jerison, H. J. 1973. *Evolution of the Brain and Intelligence*. New York: Academic Press.

Knott, C., and S. M. Kahlenberg. 2010. "Orangutans." In *Primates in Perspective*, edited by C. J. Campbell, A. Fuentes, K. C. MacKinnon, S. K. Bearder, and R. M. Stumpf, 2nd ed., pp. 313–339. Oxford: Oxford University Press.

Konner, M., and C. Worthman. 1980. "Nursing Frequency, Gonadal Function, and Birth Spacing Among !Kung Hunter-Gatherers." *Science* 207: 788–791.

Kurzban, R., and J. Weeden. 2005. "HurryDate: Mate Preferences in Action." *Evolution and Human Behavior* 26: 227–244.

———. 2007. "Do Advertised Preferences Predict the Behavior of Speed Daters?" *Personal Relationships* 14: 623–632.

Levine, A., and R.S.F. Heller. 2010. *Attached: The New Science of Adult Attachment and How It Can Help You Find—and Keep—Love*. New York: Penguin.

Levitt, S. D., and S. J. Dubner. 2005. *Freakonomics: A Rogue Economist Explores the Hidden Side of Everything*. New York: Morrow.

Lewis, R. J. 2002. "Beyond Dominance: The Importance of Leverage." *Quarterly Review of Biology* 77: 149–164.

Link, A. M. 1998. "US and Non-US Submissions: An Analysis of Reviewer Bias." *Journal of the American Medical Association* 280: 246–247.

Lloyd, M. E. 1990. "Gender Factors in Reviewer Recommendations for Manuscript Publication." *Journal of Applied Behavioral Analysis* 23: 539–543.

Maestripieri, D. 1996. "Primate Cognition and the Bared-Teeth Display: A Reevaluation of the Concept of Formal Dominance." *Journal of Comparative Psychology* 110: 402–405.

———. 2003. "Attachment." In *Primate Psychology*, edited by D. Maestripieri, pp. 108–143. Cambridge, MA: Harvard University Press.

———. 2007. *Macachiavellian Intelligence: How Rhesus Macaques and Humans Have Conquered the World*. Chicago: University of Chicago Press.

Maestripieri, D., N. M. Baran, P. Sapienza, and L. Zingales. 2010. "Between- and Within-Sex Variation in Hormonal Responses to Psychological Stress in a Large Sample of College Students." *Stress* 13: 413–424.

Manson, J. H. 1998. "Evolved Psychology in a Novel Environment: Male Macaques and the "Seniority Rule." *Human Nature* 9: 97–117.

———. 1999. "Infant Handling in Wild *Cebus capucinus*: Testing Bonds Between Females?" *Animal Behaviour* 57: 911–921.

Maynard-Smith, J. 1982. *Evolution and the Theory of Games*. Cambridge: Cambridge University Press.

Maynard-Smith, J., and D.G.C. Harper. 2003. *Animal Signals*. Oxford: Oxford University Press.

Maynard-Smith, J., and G. A. Parker. 1976. "The Logic of Asymmetric Contests." *Animal Behaviour* 24: 159–175.

Maynard-Smith, J., and G. R. Price. 1973. "The Logic of Animal Conflict." *Nature* 246: 15–18.

Metz, M., G. M. Klump, and T.W.P. Friedl. 2007. "Temporal Changes in Demand for and Supply of Nests in Red Bishops (*Euplectes orix*): Dynamics of a Biological Market." *Behavioral Ecology and Sociobiology* 61: 1369–1381.

Milinski, M., D. Semmann, and H. J. Krambeck. 2002a. "Reputation Helps Solve the 'Tragedy of the Commons.'" *Nature* 415: 424–426.

———. 2002b. "Donors to Charity Gain in Both Indirect Reciprocity and Political Reputation." *Proceedings of the Royal Society of London B* 269: 881–883.

Miller, G. F. 2009. *Spent: Sex, Evolution, and Consumer Behavior*. New York: Viking.

Mock, D. W. 2004. *More Than Kin and Less Than Kind: The Evolution of Family Conflict*. Cambridge, MA: Belknap Press of Harvard University.

Morris, R. 1996. *Partners in Power: The Clintons and Their America*. New York: Holt.

Noë, R. 2001. "Biological Markets: Partner Choice as the Driving Force Behind the Evolution of Mutualisms.: In *Economics in Nature*, edited by R. Noë, J.A.R.A.M. van Hooff, and P. Hammerstein, pp. 93–118. Cambridge: Cambridge University Press.

———. 2006. "Digging for the Roots of Trading." In *Cooperation in Primates and Humans: Mechanisms and Evolution*, edited by P. M. Kappeler and C. P. van Schaik, pp. 233–261. Cambridge: Cambridge University Press.

———. 2007. "Selection of Human Prosocial Behavior Through Partner Choice by Powerful Individuals and Institutions." *Behavioral and Brain Sciences* 30: 37–38.

Noë, R., and P. Hammerstein. 1994. "Biological Markets: Supply and Demand Determine the Effect of Partner Choice in Cooperation, Mutualism, and Mating." *Behavioral Ecology and Sociobiology* 35: 1–11.

———. 1995. "Biological Markets." *Trends in Ecology and Evolution* 10: 336–339.

Noë, R., J.A.R.A.M. van Hooff, and P. Hammerstein, eds. 2001. *Economics in Nature: Social Dilemmas, Mate Choice, and Biological Markets*. Cambridge: Cambridge University Press.

Nowak, M. A., and K. Sigmund. 1998. "Evolution of Indirect Reciprocity by Image Scoring." *Nature* 393: 573–577.

O'Brian, P. 1994. *Picasso: A Biography*. New York: Norton.

Packer, C. 1977. "Reciprocal Altruism in *Papio anubis*." *Nature* 265: 441–443.

Pawlowski, B., and R.I.M. Dunbar. 1999. "Impact of Market Value on Human Mate Choice Decisions." *Proceedings of the Royal Society of London B* 266: 281–285.

Perry, S., M. Baker, M. Fedigan, J. Gros-Louis, K. Jack, J. H. Manson, K. Pyle, and L. Rose. 2003. "Social Conventions in Wild White-Faced Capuchin Monkeys." *Current Anthropology* 44: 241–268.

Pettit, G. S., A. Bakshi, K. A. Dodge, and J. D. Coie. 1990. "The Emergence of Social Dominance in Young Boys' Play Groups: Developmental Differences and Behavior Correlates." *Developmental Psychology* 26: 1017–1025.

Piazza, J., and J. M. Bering. 2008. "Concerns About Reputation via Gossip Promote Generous Allocations in an Economic Game." *Evolution and Human Behavior* 29: 172–178.

Rege, M., and K. Telle. 2004. "The Impact of Social Approval and Framing on Cooperation in Public Goods Situations." *Journal of Public Economics* 88: 1625–1644.

Rendall, D., and A. Di Fiore. 2006. "Homoplasy, Homology, and the Perceived Special Status of Behavior in Evolution." *Journal of Human Evolution* 52: 504–521.

Rostand, E. 2003. *Cyrano de Bergerac*. New York: Penguin.

Rothwell, P. M., and C. N. Martyn. 2000. "Reproducibility of Peer Review in Clinical Neuroscience: Is Agreement Between Reviewers Any Greater Than Would Be Expected by Chance Alone?" *Brain* 123: 1964–1969.

Rowell, T. E. 1974. "The Concept of Social Dominance." *Behavioral Biology* 11: 131–154.

Sapolsky, R. M. 1992. "Cortisol Concentrations and the Social Significance of Rank Instability Among Wild Baboons." *Psychoneuroendocrinology* 17: 701–709.

Schino, G., D. Maestripieri, S. Scucchi, and P. G. Turillazzi. 1990. "Social Tension in Familiar and Unfamiliar Pairs of Long-Tailed Macaques." *Behaviour* 113: 264–272.

REFERENCES

Schino, G., S. Scucchi, D. Maestripieri, and P. G. Turillazzi. 1988. "Allogrooming as a Tension-Reduction Mechanism: A Behavioral Approach." *American Journal of Primatology* 16: 43–50.

Semmann, D., H. J. Krambeck, and M. Milinski. 2004. "Strategic Investment in Reputation." *Behavioral Ecology and Sociobiology* 56: 248–252.

Servátka, M. 2010. "Does Generosity Generate Generosity? An Experimental Study of Reputation Effects in a Dictator Game." *Journal of Socio-Economics* 39: 11–17.

Seyfarth, R. M. 1976. "Social Relationships Among Adult Female Baboons." *Animal Behaviour* 24: 917–938.

———. 1977. "A Model of Social Grooming Among Adult Female Monkeys." *Journal of Theoretical Biology* 65: 671–698.

Shubin, N. 2008. *Your Inner Fish: A Journey into the 3.5-Billion-Year History of the Human Body*. New York: Pantheon.

Simpson, M.J.A. 1973. "The Social Grooming of Male Chimpanzees." In *The Comparative Ecology and Behavior of Primates*, edited by R. P. Michael and J. H. Crook, pp. 411–505. London: Academic Press.

Skinner, B. F. 1957. *Verbal Behavior*. New York: Appleton-Century-Crofts.

Smith, J. E., S. K. Memenis, and K. E. Holekamp. 2007. "Rank-Related Partner Choice in the Fission-Fusion Society of the Spotted Hyena (*Crocuta crocuta*)." *Behavioral Ecology and Sociobiology* 61: 753–765.

Smith, J. E., K. S. Powning, S. E. Dawes, J. R. Estrada, A. L. Hopper, S. L. Piotrowski, and K. E. Holekamp. 2011. "Greetings Promote Cooperation and Reinforce Social Bonds Among Spotted Hyaenas." *Animal Behaviour* 81: 401–415.

Smuts, B. B. 2002. "Gestural Communication in Olive Baboons and Domestic Dogs." In *The Cognitive Animal*, edited by M. Bekoff, C. Allen, and G. Burghardt, pp. 301–306. Cambridge, MA: MIT Press.

Smuts, B. B., and J. M. Watanabe. 1990. "Social Relationships and Ritualized Greetings in Adult Male Baboons (*Papio cynocephalus anubis*)." *International Journal of Primatology* 11: 147–172.

Spuhler, J. N., and L. B. Jorde. 1975. "Primate Phylogeny, Ecology, and Social Behavior." *Journal of Anthropological Research* 31: 376–405.

Stein, D. M. 1984. *The Sociobiology of Infant and Adult Male Baboons*. Norwood, NJ: Ablex.

Tardif, S. D., R. L. Carson, and B. L. Gangaware. 1990. "Infant-Care Behavior of Mothers and Fathers in a Communal-Care Primate, the Cotton-Top Tamarin (*Saguinus oedipus*)." *American Journal of Primatology* 22: 73–85.

Taylor, S. E. 2002. *The Tending Instinct: How Nurturing Is Essential to Who We Are and How We Live*. New York: Holt.

Tomasello, M., M. Carpenter, J. Call, T. Behne, and H. Moll. 2005. "Understanding and Sharing Intentions: The Origins of Cultural Cognition." *Behavioral and Brain Sciences* 28: 675–691.

Tooby, J., and L. Cosmides. 1989. "Adaptation versus Phylogeny: The Role of Animal Psychology in the Study of Human Behavior." *International Journal of Comparative Psychology* 2: 175–188.

———. 1990. "The Past Explains the Present: Emotional Adaptations and the Structure of Ancestral Environments." *Ethology and Sociobiology* 11: 375–424.

———. 1992. "The Psychological Foundations of Culture." In *The Adapted Mind*, edited by J. H. Barkow, L. Cosmides, and J. Tooby, pp. 19–136. Oxford: Oxford University Press.

Trivers, R. L. 1971. "The Evolution of Reciprocal Altruism." *Quarterly Review of Biology* 46: 35–57.

———. 1985. *Social Evolution*. Menlo Park, CA: Benjamin/Cummings.

Vahed, K. 1998. "The Function of Nuptial Feeding in Insects: A Review of Empirical Studies." *Biological Reviews* 73: 43–78.

Van Hooff, J.A.R.A.M. 1972. "A Comparative Approach to the Phylogeny of Laughter and Smiling." In *Non-verbal Communication*, edited by R. A. Hinde, pp. 209–241. Cambridge: Cambridge University Press.

Van Noordwijk, M. A., and C. P. van Schaik. 1985. "Male Migration and Rank Acquisition in Wild Long-Tailed Macaques (*Macaca fascicularis*)." *Animal Behaviour* 33: 849–861.

———. 1988. "Male Careers in Sumatran Long-Tailed Macaques (*Macaca fascicularis*)." *Behaviour* 107: 24–43.

Van Schaik, C. P., S. A. Pandit, and E. R. Vogel. 2006. "Toward a General Model for Male-Male Coalitions in Primate Groups." In *Cooperation in Primates and Humans: Mechanisms and Evolution*, edited by P. M. Kappeler and C. P. van Schaik, pp. 151–172. Cambridge: Cambridge University Press.

Wegner, D. M. 2002. *The Illusion of Conscious Will*. Cambridge, MA: MIT Press.

Wenneras, C., and A. Wold. 1997. "Nepotism and Sexism in Peer-Review." *Nature* 387: 341–343.

Wheatley, B. P. 1982. "Adult Male Replacement in *Macaca fascicularis* of East Kalimantan, Indonesia." *International Journal of Primatology* 3: 203–212.

Whitham, J. C., and D. Maestripieri. 2003. "Primate Rituals: The Function of Greetings Between Male Guinea Baboons." *Ethology* 109: 847–859.

Wilson, E. O. 1975. *Sociobiology: The New Synthesis.* Cambridge, MA: Belknap Press of Harvard University.

Wrangham, R. W., and D. Peterson. 1996. *Demonic Males: Apes and the Origins of Human Violence.* Boston: Houghton Mifflin.

Zahavi, A. 1977. "The Testing of a Bond." *Animal Behavior* 25: 246–247.

———. 2003. "Indirect Selection and Individual Selection in Sociobiology: My Personal Views on Theories of Social Behaviour." *Animal Behaviour* 65: 859–863.

Zahavi, A., and A. Zahavi. 1997. *The Handicap Principle: A Missing Piece of Darwin's Puzzle.* Oxford: Oxford University Press.

Zingales, L. 2012. *A Capitalism for the People: Recapturing the Lost Genius of American Prosperity.* New York: Basic Books.

INDEX

Adam, Thomas, 218
aggression, 2–6, 14–15, 27–28, 32, 35,
	47, 103–104, 113, 140, 163, 175,
	185, 203, 217, 238, 244–245,
	247, 250, 260, 277
agonistic support/aid, 66, 203, 210
Alexander, Richard, 125
algorithms, 236–250, 266
alliances. *See* coalitions, politics
Altmann, Stuart, 36–37
altruism, altruistic behavior, 11,
	13–14, 65, 115, 120–121, 128,
	133, 151, 169, 172, 180, 182,
	213, 220, 225–227, 248, 265
analogous behavior. *See* convergent
	evolution
Andreoni, James, 121, 125
anger, 46, 136, 166, 242, 246–248,
	265
Aniston, Jennifer, 143–144, 150,
	158–169
anonymity, 113–115, 117, 121–122,
	125, 128–130, 133–141, 225, 280
anonymous peer review, 131–141
anthropocentrism, 232–234, 237
anxiety, 3, 11, 37–38, 136, 146, 166,
	193, 244–248, 264
apes, 2, 8, 36, 49, 66, 113, 155–156,
	165–166, 174, 231–232, 234,
	257–258, 262, 264, 274
	bonobos, 169, 232
	chimpanzees, 2, 18, 20–22,
		99–100, 105, 114, 155, 163, 166,
		168–169, 172, 186, 232

gibbons and siamangs, 232
	gorillas, 232
	orangutans, 156, 232
asymmetries in contests or power, 14,
	20, 28, 42–48, 51, 66
attachment, 156, 163–167, 260,
	263–264, 281
Axelrod, Robert, 12–13

babies. *See* infants
baboons, 3, 18, 29–34, 50, 99–100,
	166, 171–177, 179–180, 185–186,
	188–189, 194, 210–211, 232, 261,
	281
baby boomers, 74–75, 138
Ballmer, Steve, 89, 105
Bangkok, Thailand, 197, 199, 223
bared-teeth display, 6–7, 32, 44–45,
	114, 262–263, 277
Bateson, Melissa, 110, 113
Becker, Gary, 144–145, 154, 200
behavioral displays, 46–47, 93, 178,
	260, 278
Bellow, Adam, 53, 64, 70, 72, 74–77
Bering, Jesse, 125
Bernstein, Irwin, 38
Berry, Halle, 170
Betzig, Laura, 106
Bible, 65, 140, 236
birds, 26, 35, 113, 157, 161–163, 185,
	204–205, 207, 253, 259–260, 266
	babblers, 185
	bird nests, 205, 208
	chickens, 35

birds (*continued*)
  egrets, 26
  pair-bonding in birds, 157, 161–163, 253
  pelicans, 26
  red bishops, 207–208
Bowlby, John, 165–166
brain, 17, 37, 48–49, 83, 109, 113–115, 140, 154–155, 157–158, 162–163, 167–168, 235–237, 239, 241, 253, 256–259, 264, 266
breast-feeding, 159, 206
Bshary, Redouan, 117–118, 214–215, 217–218
Burkhardt, Richard, 230
Burnham, Terrence, 122
business partnerships, 29, 117, 127, 143–150, 169, 171, 194–196, 203, 205–206, 253, 265
Buss, David, 191, 200

Campanario, Juan Miguel, 139
Canetti, Elias, 30–31
Capizzano, Ezio, 69–70
cats, 251
Cayo Santiago, Puerto Rico, 89, 92, 101, 105, 279
cheating, 47, 109, 114, 117–119, 123, 126–128, 130, 145–147, 149–150, 171, 179, 195–196, 209, 215, 218–219, 230, 238, 248
Chiang, Yen-Sheng, 224–225
children, 9, 23–25, 27–28, 37, 46, 48, 50, 53, 60, 65, 67–69, 73–76, 105–106, 143, 145, 147, 150, 152, 155–170, 172, 183, 188, 190–191, 198, 200, 202–203, 229, 231, 237, 239–240, 247, 249, 265
Chomsky, Noam, 183
cleaner/client fish, 117–118, 123, 146, 203, 214–219
Clinton, Bill and Hillary, 151–152
Clooney, George, 96
coalitions, or alliances, 28–30, 49, 69, 80–82, 87–88, 95, 99–105, 107–

108, 115–116, 118, 171–174, 177, 186–189, 195, 202, 212, 238, 252–253, 261
commitment problem, 144–152, 156, 160, 171, 176–177, 188, 190–193, 195
communication, 6, 19, 46, 176, 178, 182, 193, 242
Conniff, Richard, 35
consciousness, 11, 39, 51, 111–114, 128, 162, 178, 234–235, 240, 244–246
contest competition, 96–98
convergent evolution, 252–253
cooperation, 5, 12–15, 22, 29, 66, 82, 108–129, 130, 132–134, 141, 145–147, 150–151, 156–157, 161, 168–169, 171–176, 185, 188–189, 194–196, 202, 204–205, 212–213, 218–227, 238, 247–248, 252–253, 260, 270, 278
cortisol, 30, 48–50
Cosmides, Leda, 241–243, 256
cost-benefit analyses, 81, 133, 141, 144–145, 147, 150, 181, 193–194, 272
Creel, Scott, 189
Crowe, Russell, 119
culture, 3, 53, 70, 147, 154, 159, 166, 193, 199, 236, 253, 262
Cyrano de Bergerac, 152–154

Darwin, Charles, 232, 236, 262
Dawkins, Richard, 232, 272
De Palma, Brian, 1
De Queiroz, Alan, 259
Descartes, 234
despotism/egalitarianism, 34, 103–105, 107–108, 227
De Waal, Frans, 37
Dickinson, Angie, 1
Dictator game, 112, 116–117, 122, 125, 224, 279
dictatorships, 72, 74, 106
Di Fiore, Anthony, 259, 261

Disney, Walt, 155, 233
divorce, 29, 143, 152, 158–160, 200
dogs, 86–88, 113–114, 189–190, 247, 251
dominance, 14, 18–52, 66–67, 79–80, 89–91, 95–108, 113, 172–175, 188, 210, 246–247, 253, 261, 278
donations to charities, 121, 124–125, 180
Dubner, Stephen, 224
Dunbar, Robin, 201

Eastwick, Paul, 163
Eastwood, Clint, 92, 191
Einstein, Albert, 231
Eisenberg, John, 260
elevator, 1–16, 244–245, 250
email, 17–20, 22, 25, 73, 86, 201–202, 250, 267–269, 271
emigration, 67, 79, 89
emotions, 37, 46, 48–49, 83, 114, 145–147, 161–162, 165, 167, 237, 240–248,
endowments, 198–200, 223
envy, 247–248
evolutionary psychology, 154, 180, 191, 200–202, 224, 236–252, 256–258, 264–265, 277, 283
eyeball-poking, 185–187
eyespot effects, 110–115, 122–123

fathers and paternal care, 67–68, 156–158, 167, 281
fear, 6, 15, 35–36, 44–47, 133, 136, 166, 241–248, 254, 258, 262, 264
fear of strangers, 166, 264
feelings. See emotions
Fessler, Daniel, 112–113, 116, 247, 265
fighting. See aggression
fish, 113, 157, 239, 252–253, 257–259, 266. See also cleaner-client fish
Fisher, Helen, 154, 159–160
fondling of genitalia. See testicles
Fraley, Chris, 157, 164, 167
Frank, Joe, 162

Frank, Robert, 145, 147–148, 150–151, 154, 160
Freakonomics, 224
free will, 232–235, 240, 283
Freud, Sigmund, 166, 237

game theory, 11–15, 39–48, 112, 116–119, 126–128, 132, 145, 195–196, 218–220, 253, 278
Gates, Bill and Melinda, 121
gazelles, 179–180
genes/genetics, 6, 14, 48–49, 65, 69, 105, 116, 154, 176, 178–179, 182, 194, 203, 227, 232, 236, 238–239, 250–252, 254–256, 260, 272–273
Gigerenzer, Gerd, 249
Gladwell, Malcolm, 224
Gore, Al, 18
gossip, 87–88, 125–127
Grafen, Alan, 182
greetings, 174–176, 188–189
grooming behavior, 6–7, 9–11, 13–15, 18, 20–22, 32–33, 38, 44, 102, 105, 173–174, 185, 203, 206–207, 209–213, 223, 277
Grossbard-Shechtman, Shoshana, 200
group selection, 227
Gumert, Michael, 206–207

Haley, Kevin, 112–113, 116, 247, 265
Hall, Edward T., 3
handicap, the Handicap Principle, 171, 176–185, 190–193, 205, 281
Hardin, Garrett, 119
Hare, Brian, 122
Harris, Neil Patrick, 170
Hauser, Marc, 128, 264
Hawk-Dove game, 39–48
Hazan, Cindy, 169
Heinroth, Oscar, 260
Heisman, Mitchell, 267–273
Higham, James, 101, 104–105
Hinde, Robert, 36–37
Hitchens, Christopher, 231

INDEX

Holekamp, Kay, 188, 212
Hollywood film industry/celebrities, 1,
    96, 143–145, 147, 150, 152,
    158–159, 169–170, 174, 197–198
hominids, 232, 256, 264
homologous behavior. *See* phylogeny
hunter-gatherers, 159

infants, or babies, 24, 67, 97, 156–159,
    165–167, 179–180, 187–188,
    210, 212, 240, 263–264, 281
infinite monkey theorem, 71
insects, 25, 35, 65, 196, 203–204,
    208–209, 214–215, 236, 253,
    258–260
    ants, 25, 65, 214–215
    butterfly larvae, 214–215
    fruit flies, 236
    bees/wasps, 25, 65, 253
    scorpion flies, 209
intervening variables, 37
Italy, 18, 53–64, 120, 129–130, 160
    divorce rates in Italy, 160
    Italian academia, 60–64
    Italian proverbs, 129–130
    military service in Italy, 53–59
    tax evasion in Italy, 120

James, William, 234
Jolie, Angelina, 96, 143–144, 150, 152,
    159, 197–198
Jorde, Lynn, 261

King, Stephen, 222
kin selection, 65, 182, 279
kinship, 13–14, 70, 77, 274
Kismet the robot, 122–123
Konner, Melvin, 159
!Kung, 159, 250
Kurzban, Robert, 202

language, 2, 24, 182, 193, 237,
    256–257, 264–265
lemurs, 3
Leone, Sergio, 92
Levitt, Steven, 224

Lewis, Rebecca, 39
Libet, Benjamin, 234–235
lizards, 260
Lorenz, Konrad, 230, 259–260
love, 2, 9, 28–29, 85, 143–170, 176,
    189–191, 193, 195, 231, 247,
    265–266, 280–281

Macachiavellian Intelligence, 223, 232,
    267
macaques, 3–7, 10–11, 14, 25, 34, 48,
    66–71, 74–75, 79–80, 89–92, 94,
    98–102, 107, 114, 128, 166, 206,
    223, 232–233, 259, 262–263, 267,
    277, 279
    Japanese macaques, 92
    long-tail macaques, 92, 94, 206
    pigtail macaques, 34
    rhesus macaques. *See* rhesus
        macaques/monkeys
mafia, 62–64, 69–70, 76–77, 278
Manson, Joseph (Joe), 98, 187–188
markets, 96–99, 102, 148–149, 177,
    181, 195–227, 248, 253, 279, 282
marriage, 28–30, 36, 44, 69, 128, 144,
    149, 151, 158–160, 171, 196–200
Martyn, Christopher, 135
matrilines, 48, 67–68, 89, 91, 188
Maynard-Smith, John, 39, 181
menstrual cycle, 168, 199, 206
Metz, Markus, 208
Microsoft, 81–82, 85–89, 95, 105, 121
Milinski, Manfred, 123–125
Miller, Geoffrey, 180
mind, 2–6, 10, 22–23, 31, 36, 49–50,
    169, 235–250, 256–257, 265,
    268, 271
mismatch, 244
Mock, Douglas, 26
monogamy. *See* pair-bonding
Monroe, Marilyn, 158–159
morality, 53, 70–71, 74, 76–77, 126–
    128, 140–141, 146–148, 150, 195,
    231, 235, 257, 264–266, 269, 272
    moralistic aggression, 127–128, 150
mothers. *See* parental care

Mother Teresa of Calcutta, 230
motivation, 44–46, 48–51, 114, 162,
    240–242, 245, 247

Nash, John Forbes, 119
Nathanson, Jared, 268–269
National Institutes of Health, 135
National Science Foundation, 134–135
natural selection, 6, 23, 39, 46–47, 65,
    147, 155, 162–163, 165, 167–169,
    177, 179, 226–227, 236–253,
    257–258, 263–264, 272
nepotism, 53–80, 138, 158, 180, 210,
    252–253, 270, 279–280
Nettle, Daniel, 110
New World monkeys, 3, 264
    capuchin monkeys, 186–187, 189
    tamarins, 156
New York City, 129, 134, 141
Noë, Ronald, 212–213, 226–227, 265

O'Brian, Patrick, 229
Old World monkeys, 232, 261–262, 264
Oprah, 194
Overbye, Dennis, 234
ovulation, 168

Packer, Craig, 29, 172
pair-bonding/monogamy, 156–170,
    247, 253, 265
parental care, 157, 163, 167, 259
partner choice, 154, 195–227
Pawlowski, Boguslav, 201
penis, 168–169, 175, 188
Perry, Susan, 186–187
personal ads, 201
personality, 82, 88, 149, 164–165
Petrie, Ragan, 121, 125
phylogeny, phylogenetic history,
    254–264
Piazza, Jared, 125
Picasso, Pablo, 229
pigs, 26
Pinker, Steven, 228
Pitt, Brad, 143–144, 152, 158, 169,
    197–198

Pleistocene, 256–257
politics, 29–30, 48, 66–69, 72–74,
    79–108, 116–117, 121, 123,
    125–128, 131, 173, 226, 230–
    231, 238, 246, 252–253, 267
power. See dominance, politics
Prisoner's Dilemma, 11–15, 112,
    116–118, 127–128, 132, 145,
    195–196, 219–220, 278
prosocial behavior, 220, 224–227
proximity, 1–16, 25, 157, 165–167,
    244, 263, 277
public goods games, 119–127, 130
punishment of defectors, 126–129, 133,
    195, 219, 280

ranks. See dominance
reciprocity, 11–14, 20–21, 30, 44, 70,
    120–125, 169, 172, 175, 182,
    213, 238, 248, 252, 280
recommendations, 53–64, 73
Rege, Mari, 122
Rendall, Drew, 259, 261
reputation, 28, 86, 115–128, 133–134,
    146, 148, 150, 191, 219, 222, 227,
    252, 279
Resource Holding Potential, 43–48,
    51, 66
rhesus macaques/monkeys, 3–7, 10–11,
    14, 25, 34, 48, 66–71, 74–75,
    79–80, 89–92, 98–102, 107, 114,
    128, 166, 206, 223, 232–233, 259,
    262–263, 267, 277, 279
rituals. See greetings
Rostand, Edmond, 152
Rothwell, Peter, 135

Sapolsky, Robert, 29–30
Schjelderup-ebbe, Thorleif, 35
scramble competition, 96–98
separation. See attachment
serotonin, 48, 83
Seyfarth, Robert, 211
sex, sexual behavior, 5, 18, 56, 64,
    66, 69–70, 94, 106, 154–156,
    162–163, 168–169, 172, 174,

sex, sexual behavior (*continued*)
180, 185, 190–191, 195, 199,
202–203, 206–207, 209, 229–231,
238, 240–241, 247, 279–280
sexual attraction, 18, 154–156, 240,
247, 280
sexual desire, 155, 162, 241, 280
sexual intercourse, 69, 156, 162, 169
sexual orgasm, 162, 169
sexual selection, 163, 178
Shakespeare, William, 29, 71
Shaver, Phil, 164, 167
Shubin, Neil, 257
Signals. *See* communication
Skinner, B. F., 183
smile, 6–7, 11, 18, 20, 44, 246, 262–263
Smith, Jennifer, 188, 212
Smuts, Barbara, 175, 189
speed-dating, 201
sperm, 205
spotted hyenas, 188, 189, 211, 212
Spuhler, John, 261
stress, 4, 6, 10, 14–15, 17, 23, 29–30,
34, 37, 44, 46, 48, 50, 55, 104,
136, 149, 165, 167, 185, 190–193
submission. *See* dominance

Taylor, Shelley, 9
taxonomy, 232, 254–256
Telle, Kjetil, 122
testicles, testis, 2, 168, 171–175, 177,
185, 192, 281
testosterone, 48–50, 83, 91, 168, 281

Tinbergen, Niko, 230, 259–260
Todd, Peter, 249
Tolstoy, Leo, 193
Tooby, John, 241–242, 243, 256
tragedy of the commons, 119
Trivers, Robert, 127
Twilight, 137

Ultimatum game, 224–225
University of Chicago, 72–73, 109,
144, 168, 200, 267

van Schaik, Carel, 93, 97, 102–103
van Noordwijk, Maria, 94
vervet monkeys, 210–212
von Fritsch, Karl, 259

Weeden, Jason, 202
Wegner, Daniel, 234–235
Wheatley, Bruce, 92–93
Whitham, Jessica, 174
Whitman, Charles, 260
Williams, Richard, 75
Wilson, Edward O., 35, 183, 270
Wimberger, Peter, 259
Wittgenstein, Ludwig, 231
Woods, Tiger, 150

Yanomamo, 250

Zahavi, Amotz, 176–185, 188–192
Zeifman, Debra, 169
Zingales, Luigi, 72